Secrets of the Forest:
Volume 1
The Magic and Mystery of Plants
and
The Lore of Survival

Secrets of the Forest Series

Written and illustrated by Mark Warren

WALDENHOUSE PUBLISHERS, INC.
WALDEN, TENNESSEE

Secrets of the Forest: Volume 1 -
The Magic and Mystery of Plants and The Lore of Survival
Copyright ©2015 Mark Bazemore Warren, 1947. All rights reserved. No part of this
book may be reproduced in any form, or by any means, electronic or mechanical,
including photocopying, recording, or any information browsing, storage, or retrieval
system, without permission from the author.
Illustrations by Mark Warren.
Photographs by Bard Wrisley, Dan McMahill, Hugh Nourse, Penny Holt, Jeff
McMahill, Adam Nash, and Susan Warren.
Type and design by Karen Paul Stone.
ISBN 13: 978-1-935186-81-6 ISBN 10: 1-935186-81-7
Library of Congress Control Number: 2016915146
   "Magic and mystery of plants; identifying and using 100 wild plants as food, med-
   icine and craft; the lore of survival; primitive shelter, water purification, tools,
   cooking, cordage, insect repellents, traps and snares. Over 200 original hands-on
   activities; 293 illustrations" -- Provided by publisher
NAT013000 Nature : Plants - Flowers
NAT026000 Nature : Plants - General
NAT034000 Nature : Plants - Trees
SCI011000   Science : Life Sciences - Botany
REF031000 REFERENCE / Survival & Emergency Preparedness
Published by Waldenhouse Publishers, Inc.
100 Clegg Street, Signal Mountain, Tennessee 37377 USA
888-222-8228 www.waldenhouse.com
Printed by We SP Corporation, Seoul Korea   11/25/2016   66902-0
Four Colour Print Group, Louisville, Kentucky

Dedications

To my mother, Marie, who gave me the freedom to go to the forest …

and

To the Cherokee, who had already marked my way

Secrets of the Forest

Learning Nature through the adventure of primitive survival skills
... and teaching it to your children

VOLUME 1.

The Magic and Mystery of Plants

Identifying and using 100 wild plants as food, medicine, and craft

and

The Lore of Survival

Primitive shelter, water purification, tools, throwing the rabbit stick,
cooking, plant foods & medicines, cordage, insect repellents, traps and snares

  over 200 original hands-on activities

Photography by Bard Wrisley
Additional photos by Dan McMahill, Hugh Nourse,
Penny Holt, Jeff McMahill, Adam Nash, and Susan Warren
Thanks to Darryl Patton, publisher of *The Southern Herbalist & Stalking the Wild* for re-
viewing the medicinal/edible plant uses of survival scenarios included in this volume.

The SECRETS OF THE FOREST Series
~ The Books of Medicine Bow ~

Volume 1:

<u>The Magic and Mystery of Plants</u> (Identifying and using wild plants for food, medicine, and craft) *and* <u>The Lore of Survival</u> (shelter, water, rabbit stick, hunting, tools, insect repellents, cooking, traps & snares, food, and medicines)

Volume 2:

<u>Calling up the Flame</u> (Everything you need to know about FIRE: pyre-building, wood selection, being match-savvy, sustaining a fire, fire-by-friction using the hand drill, bow drill, and fire saw) *and* <u>Feeding the Spirit</u> (ceremony and storytelling with purpose)

Volume 3:

<u>Eye to Eye with the Animals of the Wild</u> (Stalking, tracking, hide-tanning, and snake lore) *and* <u>At Play in the Wild</u> (games: academic, adventurous, and around the fire)

Volume 4:

<u>Projectiles</u> (Making bows and arrows, the art of archery, throwing spear, knife & tomahawk; the atlatl) *and* <u>The Blessed Path of Water</u> (whitewater canoeing: from lake to river)

TABLE OF CONTENTS

Foreword

The book *Last Child in the Woods* by Richard Louv laments that our children no longer wander in the forest, or know where their food comes from. This disconnect from their true home, from the treasures of the world around them, from their backyard, is not just a casual loss.

Mark Warren has spent his life welcoming children back home. A six-year-old child asked me in a school hallway, "Is Mr. Mark coming today? I brought him a plant." For 40 years I have watched this lover of the earth work his magic with students.

These four volumes allow us, whenever we turn a page, to allow Mr. Mark to arrive, to learn what plants the Native Americans used for medicine, how they taught their children to stalk animals, how to use a bow, or to throw a knife , or tell a story.

All the richness we are homesick for – the questions we were so curious about as children – echo in these books. You will treasure these volumes. The heart of a child that lives in each of us will resonate with this magical man's journey. Take a walk with him into a world we've lost.

Herbert Barks

Herbert B. Barks has been headmaster of the Baylor School in Chattanooga, Tennessee and the Hammond School in Columbia, South Carolina, each for 17 years. His commitment to experiential education created the Walkabout program, one of the premier wilderness programs in the nation. Mark Warren worked for 15 years with this program as it was formulated.

"In the old days children soaked up the colors of the world through the windows of their eyes. We gave them chores and they worked with their hands. We talked to one another at meals and in the evenings. Now our children are looking into bright, little windows that we have handed to them – computers and cell phones and electronic games. They've looked away from the world."

~ *a summer camper's grandfather*

Author's Note

Survival skills can be divided into two camps: primitive and high-tech … or put another way, Native American survival and military survival. This series of books is about the former, which, I believe, ultimately must rate as the superior approach for one simple fact: its resources come not from stores but from the forest and are never depleted … unless grossly over-harvested.

In high-tech survival, special tools and chemicals are parts of the survival kit. Each and every piece of that kit – whether it be matches or a water filter or a hunting rifle – has the potential to fail or expire. Then what? Take the gun, for example: Unless one has the chemical means to forge metal and produce gunpowder indefinitely, there must come a day when the ammunition is gone. Not so with a bow. A primitive survivalist can always make a new bow to replace a broken one. And the arrow-maker's trees, shrubs, brakes of rivercane, feathers, and stone will always be out there waiting.

In a very real sense, learning the primitive survival skills of a given eco-system is tantamount to studying the native people who once thrived there. In the developmental years of civilization it was the land that shaped cultures, and today anyone who re-approaches the act of "living off the land" bumps into the same challenges that the first people of that geography encountered. The resources from which a "modern primitive-survivor" might draw a solution to a specific problem are, in most cases, the same resources that were used by ancient people.

For example, any person of any era who immerses himself into woods-lore quickly learns the value of cordage. As tool-makers we have the need to attach things to one another … to fasten together the parts of a tool or structure. Lashing with cordage is a fundamental way to do that. The modern woods-person is eventually going to experiment with plant fibers that can be twisted or woven into rope. (Of course, there are animal parts that provide cordage, too, but plant fibers are especially appreciated for their ease in harvesting. Plants don't run away from us.) As the contemporary student experiments and strips fibers from weed stems, leaves, roots, and the inner bark from various trees and shrubs, he is following in the moccasin prints of history.

The Cherokee lived (and many continue to live) in these mountains of Southern Appalachia that I call "home," and so my personal quest to learn the plants

and animals and the skills of primitive survival necessarily placed me on a path of Cherokee study – an experiential education that led to an academic devouring of books on Native American life. It was only natural that I would want to see how my unseen "mentors" lived in this place … in this same forest.

My place is, in fact, called *Medicine Bow*, a wilderness school in the Southern Appalachians of north Georgia. As a pre-med student who majored in chemistry I was introduced to the chemicals assembled inside the miraculous world of green leaves. Later, as a teacher of Nature, I have been privileged to carry this information to many thousands of students from preschooler to octogenarian. Over a half-century of teaching I have developed Nature activities, games, and lessons designed to make the learning fun. I use the same programs for all ages with some simple modifications to accommodate each level of maturity. I have compiled it all in this series of books so that I may pass it on to you and to the future. This book is like a child forever up for adoption, never growing old, hopefully enhancing and enriching whoever will foster its lessons. These lessons are a thousand airborne seeds and you are the wind. My wish for you is that you approach these ancient skills with a primordial hunger and a sense of adventure and that those ingredients become so integrated into your recipe for learning that they infect everyone around you.

This book shows you how to explore "the real world," to know it as your home. That "real world" – whether it be forest, field, swamp, prairie, or desert – is waiting for you to return to your primitive roots. No matter your nationality, you are descended from ancestors who practiced the ancient skills of survival, and in embracing this adventure you are reentering the same kinds of wild places that were known so intimately to your paleo-kin. True, these natural places might now be modified by human encroachment or by alien plants that have found their way into your area, but – with some exceptions – it is roughly the same historic biodiversity of flora and fauna. The stage is set to pluck the atavistic string deep inside your DNA.

As a suburban baby-boomer I had some advantages not so available today. In a world less crowded, there were wooded lots scattered throughout my neighborhood. Without computers and videos and electronic games (and with a one-TV-show-a-week mandate from my parents), I had *time* and a *reason to go outside*. From an early age I had neighborhood comrades, and whenever we burst from our houses we let loose a high-pitched pre-pubescent trademark signal that carried a half-mile over the low hills – a looping sound with a sharp upward end-note that gave the call its question mark. (We couldn't yet whistle.) This signal was a call to assembly. Basically it said: *I'm free! Ready for action! Anybody else out here?* This call always heralded the beginning or continuation of an adventure: pole vaulting over the gully, building a tree house, carving model boats with scrap-lumber and running them on epic journeys down the creek.

Without the fear of the multitude of crimes that dominate the news today, I had the freedom to explore. My mother expected nothing more from me than to return for supper. The appetite worked up by such an adventure never failed to deliver me back home.

Despite the blessing of having good friends, I spent most of my time in the woods, by choice, alone. For a young boy, an undeveloped block of forest was a vast territory, and by being in it, I was reaching – like Adam's hand on the ceiling of the Sistine Chapel – feeling for the electric tingle of the Creator's touch. Even though I did not know the names of the majority of plants around me – nor could I identify most of the tracks pressed into creek sand – I knew that I was in good company. I was standing in the flow of things honest and vital and filled with secret purpose. Nature was ancient and complete and so much bigger than I. And for reasons I could not yet articulate, I needed to be a part of it. In the shadows of that forest, I *thought* I had been alone. When I grew to be a man, I came to understand that the trees had known me all along … literally.

In this present age of computers, smart phones, iPads, and other electronic gadgets (and the sedentary lifestyles that accompany their uses), education analysts now point to a lack of contact with Nature and the disappearance of "unstructured play" as a void that saps our children of their innate vigor and physical health. Perhaps more importantly it denies them any sense of stewardship toward the world that they will inherit. Richard Louv, author of *Last Child in the Woods*, has aptly named this malady "Nature-Deficit Disorder."

Many scientists believe that we have reached a critical time in the history of human habitation on the Earth. Some think we have taken the planet to a state of such degradation that the harm cannot be repaired … the trend not reversed. Let's hope not.

What we need is hope. If we are to have hope we must produce a generation that is enlightened about healing the environment. But first – before children will see a reason to protect Nature – they must enjoy it. They must spend time in it to learn how to appreciate it. They need to eat a Solomon's seal tuber, stalk a deer, rub pawpaw leaves onto their skin to repel gnats and chiggers and mosquitoes. To experience all this, they have to get out there … with the people charged with their education. If parents and teachers do not lead our computer-dazzled children into the forest, who will?

In her work, *The Sense of Wonder*, Rachel Carson wrote about an intimidation factor that prevents parents from exploring Nature with their children. The adults feel inadequate. The physical world seems so complex and unfamiliar that to teach it seems a hopeless prospect.

To that conundrum I propose this recipe: Choose one survival skill in this book that interests you or your child, and then get out there in the real world and meet it head-on. Stumble into it on your own, if you wish, as if you are the first paleo-humans to address the subject. Get your hands dirty. Explore. (There is one exception: **Do not experiment with wild foods and medicines! For that we need academic guidance to compensate for our lost instincts.**) Then read the text together and try the accompanying activities. Before you know it, you're on your way.

PART 1

The Magic and Mystery of Plants

"Every plant you see around you has a God-given purpose ... some to cure a disease, others to keep off the skeeters, and plenty to nourish you. If people tell you a plant has no value, it just means that they have forgot the old ways."

Jonas Walks Through the Storm, Song of the Horseman

CHAPTER 1
In the Beginning ... There Are Plants

To make a good bow, one that will cast an arrow with authority time after time, a bowyer must know how to recognize trees in winter (without their telltale leaves), because winter is the preferred time to cut a tree for a bow. And not just any species of tree will do.

To create a flame a fire-maker must be selective not only about the condition of the pieces of dead wood that he will spin one upon the other ... but also about its identity. He must know the dead material by name, because – again – not just any species will do.

In constructing his pyre he'll want, as the framework of his structure, sticks that do not quickly ignite. As kindling, he'll want wood that burns aggressively by its volatile chemical content. Then, larger porous sticks that dry quickly. Finally he transitions to slower burning hardwoods that make lasting coals. All of these needs are met by an intimate familiarity with plants ... dead ones, at that.

There is a tree whose inner bark resolves the pain of a common headache. Another a migraine. One plant's root dispels nausea. From another tree, a leaf's scent wards off insect pests. Out in the wild are trees, shrubs, vines and herbaceous plants that contain expectorants, basketry strips, fish poisons, edible tubers, styptics, poultices for stings ... the list goes on and on.

Though our remaining American forests may be somewhat changed from pre-Columbian days, foods, medicines, and craft materials still abound out there just as they did when the Native Americans met all their needs through Nature's gifts. Wild land still stands today as a world complete. It could be in the form of a mountain range or a mere wooded lot situated somewhere in your neighborhood.

Yet today – beyond providing a postcard backdrop of eye-pleasing beauty – this superstore of gifts remains largely an unknown place in the American culture. This nurturing green world is quietly, patiently awaiting recognition by anyone who would seek it.

Learning about plants – their names, relatives, scents, chemicals, nutritional value, and craft uses – weaves together the complicated tapestry of survival skills. The more we learn about plants, the more we begin to understand animals, terrain, geology, and ecosystems. Animals of the wild have preferred plant foods. These plants have preferred growing locales. Therefore, to find them, we should seek them out in those logical habitats.

> *Plant study is the foundation upon which all survival skills are built.*

Understanding plants helps us to become better trackers, stalkers, tool-makers, and shelter builders. Plants teach us how to be comfortable in the wild by using their foods, medicines, insulation, fuel, fire-making tools, and insect repellents. Plant study is the foundation upon which all survival skills are built.

The Relationship Between Plants and Humans

Anthropologists teach us that primitive people enjoyed the same kinds of instincts about plants as wild animals do to this day. Imagine a Stone Age cave-dweller who felt a headache coming on. Inexplicably his feet began to lead him to lower ground. At a stream he was drawn to a stand of willows, where he stripped away leaves and nibbled on a tender green branch shoot to fill his mouth with its bitter juice. Despite the taste, he swallowed the liquid and continued to chew. In time the headache disappeared. The elements of this scenario might show the same lack of fore-thought and planning as on a cold day when you rub your hands briskly to warm them. Who taught you to do that? Who teaches a rat snake (which hatches from a cached egg and never meets its parents) how to hunt? Which animals should it eat or avoid? Who teaches a bird to fly?

Instincts are hard-wired into hereditary material, but they can be lost. Where plant usage is concerned, almost all humans today lack the instincts that our ancestors once enjoyed. Because of that, we must approach a relationship with plants on an academic level. And we must do it unerringly, for this is one area where a mistake can be costly.

Many plants make formidable poisons for self- protection. Some are lethal to humans. Each year people die from ingesting the wrong wild plants, the wrong parts of plants, or plants out of season. Often the tragedy falls to unsupervised children.

The efficient use of plants is one of the best magic wands that a teacher can wave in converting students from an audience-mentality into a state of experiential interaction with *the real world*. People of all ages are curious to taste a wild food … or to rub a juicy stem into a skin rash to feel a bothersome itch abate … or to bruise a leaf and wear its scent to repel mosquitoes.

Finding immediate solutions to our problems by using wild plants fascinates us. What today seems esoteric was once the daily fare of all men, women, and chil-

dren. Just as it was for the ancient ones, this lore remains *a gift* for those who embrace it today. A conscientious student who has benefited from a medicinal plant can repeat the experience *if* she is willing to spend enough time with the plant so as to positively identify it again at another place at a future time. This is not merely data cluttering the brain. It is practical knowledge cemented into memory with purpose. It is experiential.

In one way plant study has been made easy for us. Plants are not like deer and fox – those phantoms of the forest that dissolve into the shadows at our approach. Plants wait for us.

It is this simple "rooted" living style that defines the true miracle of green plants. They can "settle" at a given location because of a remarkable capacity to **manufacture food**. While the fauna of the forest moves about through their waking hours in a continuous search for food, the green plants soak up their needs from dirt, water, air and sun … and create their own food.

Within this process, green leaves extract packages of energy from sunlight and store it in the sap they make. This is the energy that runs the world – from the twitch of a mouse's tail to the roar of a jet engine. Green plants are the "middlemen" in this energy-transaction, serving as agents for the rest of us, trapping solar energy on/for this planet. We humans are as much a part of that equation as any other living creature. We obtain that solar energy by eating plants, animals that eat plants, or animals that eat animals that eat plants. Whenever we sit down to

> *Plants are not like deer and fox – those phantoms of the forest that dissolve into the shadows at our approach.*
>
> *Plants wait for us.*

the Standing People

a meal we should remember that the plate of food before us is actually an array of concoctions of sunlight catered to us by a host of leafy green chefs.

The Cherokees called trees "the Standing People." Most researchers might assume that this modifier – "standing" – was chosen simply because trees *stand* so tall. I propose a more profound meaning. The trees are "standing" as opposed to "mobile." They don't need to move around. They have cornered a significant shortcut in the survival equation. Rather than forage or hunt, they construct what they need, and in the process capture energy from the sun.

The story has a mythological feel – like the tale of an epic hero who steals fire from the gods and then shares it with the shivering masses huddled on the Earth. Understanding this gift, how is it possible for us to look at the Standing People (and all other green plants) without seeing them as our allies in the survival equation? Though the first people could not have known the specifics of photosynthesis, I am convinced that they understood some version of it. I believe the Cherokees chose their adjective well, based upon the miracle of greenness.

We, of course, continue to depend upon green plants. Without them there could be no life on Earth. Yet we don't seem to behave accordingly. Most humans probably go through any given day without acknowledging the essential gift of greenery. We raze whole sections of forest without a thought of loss. We spread plant poisons as if we have been forced into warfare. How are we able to do that? This is an important question to ask, because our ability to educate our children about Nature hinges upon our understanding of this basic relationship between plants and humans.

Perhaps such aloofness creeps into our lives because, in our modern American culture, it is possible to go about our days without our feet ever leaving pavement or carpet or flooring. Weeks can pass without our fingers handling a leaf or a stick or a stone. For much of our lives Nature is hidden from us, shielded from our daily experience by walls and thermostats and grocery stores. We have invented a way to live blinded to our true benefactors, when in fact every particle of Nature around us – stone, water, insect, weed, and wind – nurtures us like a devoted mother who seems to love us no matter how we behave.

In harvesting a part of a plant for food, medicine, or craft, the Cherokees moved through a series of rituals that included circling a plant, approaching it from the south, speaking to the plant, and perhaps bestowing a gift of thanks. Certainly this must have prompted a private chuckle from the Europeans who first came to the Americas. They would not have understood such interaction. But now we can presume that those primitive people – so interwoven into their natural surroundings – knew what they were doing. On a gut level they knew what our scientists are only now uncovering about the lives of plants. Plants have awareness. Plants respond to what is around them – even people. And plants communicate. Plants have nervous systems based not solely upon turgidity (inflation or limpness based upon water pressure), as was once believed, but upon electrical pathways, not unlike our own nervous systems.

Plants communicate with one another using chemicals called *pheromones* – scents that they manufacture and let waft into the air. In 1983 researchers at the University of Washington and at Dartmouth verified this by studying two like trees, one encased in a mesh net into which researchers introduced ravenous caterpillars that began to devour the tree's greenery. On the other tree – the one untouched by predators – leaves began a buildup of chemicals that proved to be offensive to the invaders. How so? An invisible emergency message had been sent from the besieged tree as a warning to other trees. As a result any tree receiving the message armed itself with increased insecticidal properties, thereby rendering its leaves unpalatable.

Animals (including humans) make pheromones, too. Is it possible that there could be communication between humans and plants? Could those plant pheromones breach the barrier between plants and humans? Why do we presume there is a barrier?

Pheromones are well documented in humans, though we have probably seen only the tip of the pheromonal iceberg. Women stranded in a cabin for a time discover that their menstrual cycles synchronize. The same thing happens in female-dominated office settings. "Sexual" pheromones have been identified on humans around both genitalia and armpits. This explains the spongy retentive texture of hair tufts located there. No matter how TV commercials try to brainwash us about this idea, that natural scent is part of our sexual identity. (One wonders if our current American diets and obesity levels might render us less attractive compared to paleo-standards.)

Along with facial expressions, posturing, hand gestures, and crude vocalizations, pheromones served as an early way to broadcast human emotions. Perhaps there were chemical scents for dominance, jealousy, compassion, anger, and a host of other common reactions. Perhaps there still are. But our pheromonal acuity has been dulled, if not lost. How did this happen? Undoubtedly, the culprit was one of our most beautiful achievements: the refinement of vocal language.

Pheromone exchange seems to be what the Native Americans were practicing with plants. They knew something about interaction with Nature that we have unearthed by scientific experimentation only over the last decades. Was there a practical basis for the Cherokee ritual during plant harvesting? What if a plant changed its chemistry in the presence of threatening pheromones? What if the way you behaved with a plant at the time of harvest dictated its efficacy as a medicine or food? Some researchers suggest that this is why the Native Americans used sacred formulas and rituals … such as speaking to plants. For "primitive" people this was likely an innate courtesy based upon gratitude, not unlike the saying of grace at a modern dinner table. Is it not appropriate to speak to the Creator as we gaze upon the countenance of a living plant?

What if the way you behaved with a plant at the time of harvest dictated its efficacy as a medicine or food?

"To know a plant well, you must sit with it for as long as it takes to draw its picture. The next time you see that plant in another place, you will encounter an old friend."

Robert Spotted Horse, A Copperhead Summer

CHAPTER 2
The Pieces of the Whole

Entering into a study of the plants that grow in your immediate environment is an eye-opening adventure. It is like waking up one morning in the same house where you have been sleeping for years and discovering that someone had long ago invested a lot of thoughtful time and energy into the working parts and interior decoration of that house. Like all the other survival skills, plant study is a fulfilling journey when you have the luxury of learning it at your own leisure, not pressed into it with desperation in a dire situation. So let's have fun with it. All the forest knowledge that you amass now will be invaluable not only in an emergency but also in the simple joy of "returning" to the wild.

To master plant identification you must first learn how to look at a plant. It is similar to picking apart the components of a symphony. As a child you heard the whole of a composition – a gestalt approach – but as you became aware of different instruments, you learned to recognize them, to single them out. This chapter will teach you to do that with plants – to isolate their anatomical parts, their colors, shapes, textures, and scents. The new words that you will learn comprise a veritable "language of botany." With this you will learn how to communicate your findings to someone else. More importantly, by knowing this language, you can tap into the wealth of information that botanists have passed down to us through the books they have written.

The academic study of plants is vital for your grasp of primitive skills, for it lays the groundwork for all that will follow. Gather your child to you on a rainy day and nestle up to the table with paper and pencil and put in the time that this chapter asks of you. Later it will reward you many times over as you confront the physical challenges of each individual survival skill.

The Self-made Botany Book

When I visit a school classroom – and this could be any grade level from second grade to senior year in college – to present a botany program, I guide the students

through the making of personal booklets for plant study. They take six unlined sheets of paper and fold them into an 8½" X 5½" booklet, then at the top of the front cover they write "BOTANY," and then "WRITTEN AND ILLUSTRATED BY" … and finally their names. Younger children enjoy binding their pages with a piece of yarn. Most of the cover is reserved for a drawing that will purposely be rendered at a later time. (A drawing of what? I don't tell.) Right away the students have a sense of belonging to this study, because they participate in the making of the booklet. Instead of being on the receiving end of a lesson … they are on the creative end.

Make this booklet for yourself. This chapter will lead you through the process. On the pages of your booklet, draw all the pictures that you encounter here. There is no better way to know the physicality of a thing than to draw a picture of it. Sketching something elevates your eye to new heights of "seeing." It forces you to examine.

There will be eleven chapters in your self-made book, most of them consuming one page each. Be concise. Make good use of the space on the paper. After all, now that you are a budding botanist who is headed toward using the raw materials that plants offer, you will become more aware of plant products. Take a moment to touch the paper, to remember that this is the flesh of a tree. Our minds know it, but our sensibilities forget. We handle trees everyday: newspapers, books, magazines, tissue, and the biggest pulp product of all – cardboard.

As the students open their self-made books to take notes, I watch to see on which side of the booklet – recto or verso (right or left page of an open book) – their pencils hover. Suggestions are sometimes in order. (Why start off wasting a verso?) I ask them to write these words on the left page at the top of the paper: "An Uneducated Drawing."

An Uneducated Drawing –

Go outside or visit a potted plant inside. Select a simple plant less than one foot in height, sit down next to it, and get comfortable. Render its portrait inside the front cover of your booklet on the verso page. Work independently in a private space. The purpose of this exercise will be explained at the end of this chapter.

Creating the Botany Booklet –
Your first step on the path to know plants

Botany Booklet Chapter 1 – Leaf Shape

On the first *recto* page inside the booklet, draw the three simple leaf outlines shown below, giving each a **leaf stalk** so that the leaf's **base** and **tip** are defined. Make sure that the first leaf is widest closer to the base of the leaf, the second widest in the middle, and the third widest nearer the tip. Label the parts as shown. These leaves differ only by where that widest span is located on the blade.

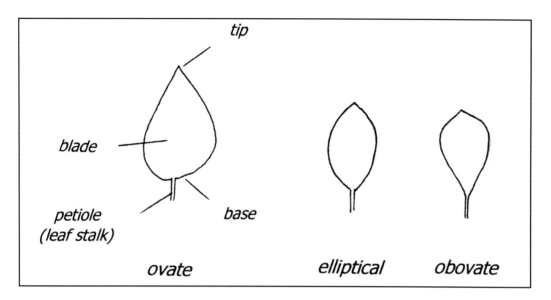

Here's an observation test to see if you've been paying attention all these years as you've walked past countless species of plants: Which leaf design is most common in your area?

In most parts of North America the first drawing is the correct answer, followed by the second drawing, and then the third. Now let's give these shapes a name.

Botany, like the other sciences, utilizes Latin for its naming and organization. The Latin word for "eggs" is *ova*. Because the first drawing reminded a scientist of the silhouette of a hen's egg, a leaf with its widest part closer to the base is called **"ovate"** (making an adjective out of "*ova*"), which means "egg-shaped." The second, more symmetrical leaf is called **"elliptical."** Since the prefix "ob" means "opposite," the third leaf is called **"obovate"** – its linear symmetry being the opposite of the first.

There are many more leaf shapes than these. Draw a longer, thinner blade (say 5" long and 1" wide) being careful to have the widest portion closer to the base. This design reminded a scientist of the point of a spear, and so it is called **"lanceolate."**

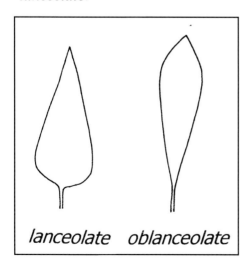

Now draw a similar blade but have its widest part nearer the tip. Can you name this?

There are leaves more complicated than these few examples, some with appendages that jut out from the main body of the blade. Each design has its own name. For example, a star-shaped leaf is called **"stellate."** A heart-shaped leaf is called **"cordate."**

At this point it is not necessary for us to learn every shape and its name. You'll

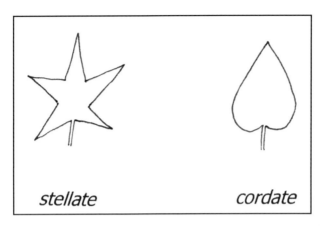

stellate   cordate

accrue that information over time, learning the names of new shapes as you encounter them outside. For now you can refer to these leaves in more general terms. Calling a leaf "star-shaped" or "heart-shaped" or "arrowhead-shaped" is acceptable; in fact, many plant books written for the layperson use such generalized terminology.

But you *do* need to know what to call those leaf blade appendages and the empty spaces between them. The appendage is a **lobe**. The space between lobes is a *sinus*.

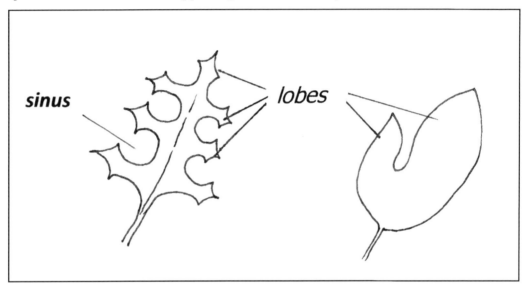

sinus   lobes

Drawing What Isn't There –

Outside, choose a lobed leaf that faces you and draw everything you see around it in the background. Pay special attention to what you see through each sinus. The resultant drawing should leave a blank white space that perfectly silhouettes your chosen leaf.

In the field, whenever considering leaf shape (or any other aspect of a plant), always examine several different leaves on the same plant. Make a determination of the ratio of leaf length to width. Is the leaf twice as long as wide? Three times as long as wide? This observation helps you to render an accurate drawing. Such attention to dimension makes the drawing true to its subject.

A Survey of Leaf Shapes –

Go outside and take a walk along a random path to take a survey of any and all plants that have leaves *without lobes* to see what percentage of these plants have

ovate leaves. Do the same for lanceolate, elliptical and all the other leaf shapes you have learned up to this point. Besides making a record of your natural surroundings, this is way to introduce a math class to ratios, percentages, graphs, and charts.

Look again at the *base* of a leaf. It can flare slowly or abruptly from the leaf stalk. Perhaps it forms the V-notch of a heart. Such a base can be symmetrical or asymmetrical.

The leaf *tip* has its nuances, too. How do the sides converge toward the tip – abruptly, gradually or on an S-curve? There are botanical words for all these variations. Any plant ID book will define them for you in its glossary. For now, the generalized descriptions introduced thus far will work nicely.

Botany Booklet Chapter 2 – Leaf Margin

All the leaves we have drawn so far have margins (edges) easily sketched with a gliding pencil stroke (complicated only by lobes, if present). That's because they all possess **smooth** margins. Draw that first ovate leaf again; but, this time, render its margin jagged like the teeth of a saw. This margin is called "*serrate*" or "*toothed*." If a leaf has more than one size of tooth, it is **double-serrate** or **double-toothed**. Rounded teeth can be **crenate** or **undulate** depending on the severity of the curve between the teeth.

A close inspection of a seemingly smooth margin might reveal a fringe of tiny hairs making the margin **ciliate.**

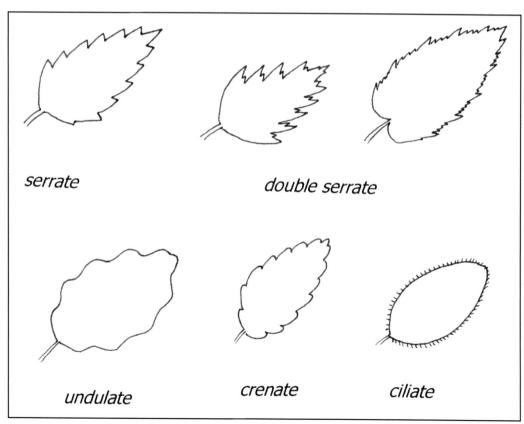

serrate double serrate

undulate crenate ciliate

 <u>**A Survey of Margins**</u> – Walk again on an arbitrary path and take notes on the numbers of plants with smooth, serrate and double-serrate leaves to get a feel for their distribution in your area. This time make a bar graph or a pie chart of your findings.

Now that you know about the possibilities for lobes and margins, you are ready for a common term used by botanists to describe the simplest leaf shape of all. If a leaf has a smooth margin and no lobes, it is called *entire*. The first five leaves in your booklet are examples of "entire."

Botany Booklet Chapter 3 – Leaf Attachment

 Some leaves have no stalk attaching them to the stem (trunk) or to branching (limbs, twigs). If a leaf emerges directly from the wall of the stem or branch, such a leaf is **sessile** or **stalk-less**.

All leaves we have drawn thus far do have stalks. (We added those stalks simply to help define the base and tip for each leaf.) Such leaves are aptly named "**stalked**." Leaf stalks (also called "**petioles**") come in many lengths. This length is noteworthy. In your drawings make the scale of length-of-petiole to the length-of-leaf an accurate one. Use a broken twig as a measuring device. Sometimes a plant produces varying lengths of petioles for leaves positioned at different heights on the stem. Include this in your drawings.

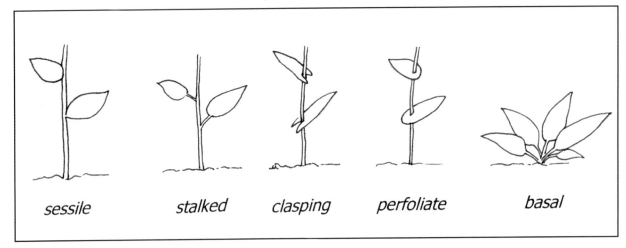

sessile   stalked   clasping   perfoliate   basal

Note: The word "**stalk**" is often erroneously used in the place of "**stem.**" As botanists, we need to know the difference. The stem is the main trunk of a plant. A stalk attaches a bud, leaf, flower, or fruit to a stem or branch.

If a leaf appears to "pinch" the stem, it is called "**clasping.**" Either the blade partially wraps around the stem or the petiole is tube shaped, enclosing the stem like a slitted sleeve.

<u>Undressing Grass</u> - Find a stem of grass long enough to have several blades (leaves). Choose one blade, pinch it between thumb and finger and gently pull the clasping part of the leaf stalk away from the stem. Then, to test your "surgical" finesse, slip the clasping sleeve back onto the stem just as you had found it.

One leaf attachment surprises students when I draw it, but a **perfoliate** attachment is not rare. When I walk students to a specimen of *perfoliate bellwort* or *boneset*, they are twice surprised: once at the structure itself and again that they have been walking past such a unique design without noticing it. *Perfoliate* comes from the Latin phrase *per foliatum* (meaning: "through the leaf"), so named because it looks as though a stalk-less leaf has been impaled by the stem.

A fifth option exists – one that is good to introduce as a think-outside-the-box riddle. Ask your students: "Can anyone visualize a plant that has multiple leaves but no stem?" In a classroom it's fun to have the students come to the board to draw from their imagination. There are usually a few false starts until someone gets it right. A dandelion might be the best known example.

When leaves emerge directly from the root at ground level with no stem rising from that root, the leaves are called **basal**. (Not to be confused with the mint named "basil." In fact, *basil* is not *basal*.) A cluster of basal leaves that radiate in all directions is called a **basal rosette**. An evening primrose, for example, is easy to identify when in flower or in fruit, but to harvest its young root for a wild food, one must learn to identify its first-year basal rosette. Because primrose is a **biennial** (it lives for two years), its stem develops in the second year.

perfoliate bellwort

<u>A Survey of Attachments</u> – Take a walk outside to determine the most common leaf attachments in your area. Be sure to gently tug on leaves to see if they possess hidden clasping petioles, just as grasses do.

Botany Booklet Chapter 4 – Leaf Arrangement

Draw all four types of leaf arrangements shown below. Which is most common in your area?

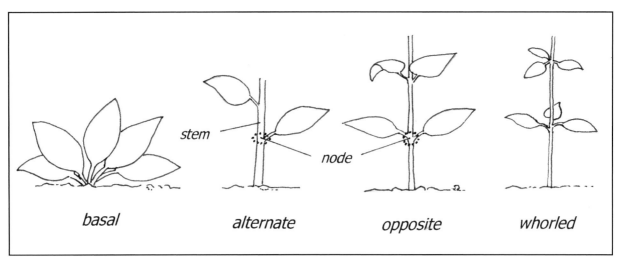

basal   alternate   opposite   whorled

A **node** is a place on a stem (or branch) from which a leaf emerges. Sometimes nodes are swollen, sometimes not. Think of it as a joint of the stem or branch, because zigzagging plants zig or zag at the nodes. Sometimes a node shows nothing remarkable. It is merely the height on the stem associated with its leaf. This term "node" will help us define leaf arrangements.

A plant with **opposite** leaves (and, therefore, opposite branching) has 2 leaves per node. Opposite leaves do not necessarily emerge from opposite sides of the stem (180° apart), but they usually do. **Alternate** leaves show 1 leaf per node. (This arrangement is most common in the American wild.) **Whorled** leaves show more than two leaves per node. **Basal** leaves have no stem.

Knowing the handful of opposite-leaved trees is very useful in the identifying process. This list includes **maple, ash, buckeye, viburnum, bladdernut, princess tree** (an import spreading prolifically) and all but one of the **dogwoods**. Opposite-leaved *shrubs* are more numerous, and there are a great many more opposite-leaved *herbs*. (*Herbs* are not merely those flavor enhancers used in your kitchen but any of the green, non-woody plants of forest and field that flower. Many people refer to herbs as "wildflowers" or "weeds.")

Though we will explore this in more detail in the chapter on winter botany, be aware that there is a "spiral" theme involved in leaf arrangement (and in many other aspects of plant growth). In other words, alternate-leaved plants do not necessarily alternate their leaves 180°; that is, you don't find one leaf on the "left" side of the stem (or branch) then the next leaf (at the next node) on the "right." If you draw parallel stripes longitudinally along the stem through each spot where leaves emerge from the stem (or branch), you might find that the spacing between stripes is 120° or 144° or some other portion of the available 360°.

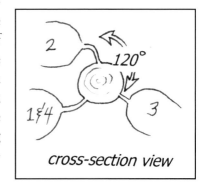

cross-section view

Each plant species adheres to its own formula, although the measurement can sometimes seem inexact because of natural deviations caused by any number of factors, such as: animal predation, severe weather, or the torque of a stem or

branch as it grows. There are plants (birch, basswood, elm) that do alternate leaves in a true left-right alignment (180°) giving their branches a fan-like appearance since all the leaves of one branch are flattened in the same plane.

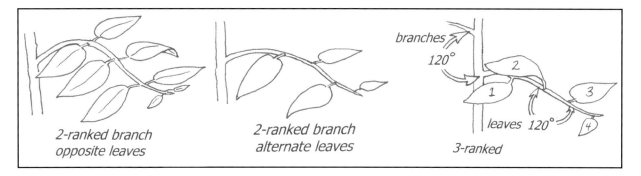

2-ranked branch
opposite leaves

2-ranked branch
alternate leaves

3-ranked

Each one of these parallel stripes imagined along the stem is called a "**rank**." To understand the concept of ranks, consider a plant in the mint family, which grows its leaves in an opposite arrangement. The mint stem is conveniently 4-sided, making a square cross-section with four corners. If you were lying on the ground studying a mint at eye-level, let's call the side of the stem facing you "side A", the next side to your right is "B", the side away from you "C" and the one to your left "D."

If the lowest pair of opposite leaves (1 and 2) fan out to either side from stem-sides B and D, looking up the stem to the next node you will likely find the next pair of leaves (3 and 4) turned 90° from the pair below. One leaf points toward you and one away, these leaves being on stem-sides A and C. The next node has a pair of leaves (5 and 6) oriented like the first pair on sides B and D. How many ranks are present? With a square stem, the answer is easy: There are 4 ranks (A, B, C, and D). If this stem were round in cross-section it would still be four-ranked.

Looking at leaf arrangements on trees in the spring can be deceptive to the eye, because as buds open, not only do leaves emerge but also new twig tissue emerges. This nascent branch is visible to the eye and ready to elongate, but so early in the growth period you'll find alternate leaves may be bunched

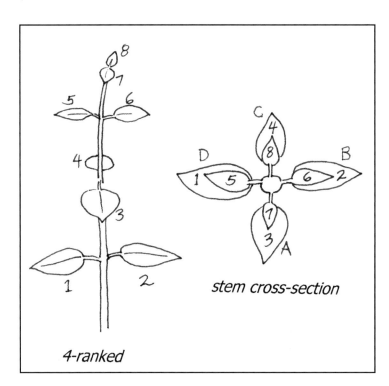

stem cross-section

4-ranked

birch spurs

together so as to appear opposite or even whorled. It takes an exacting eye to differentiate it. In spring and early summer look at many different places on the tree to discern leaf arrangement.

Some trees (black birch and sweetgum are examples) are prolific producers of dwarf twigs called **spurs**. Spurs help to fill in the inner space of the tree's canopy. They are so stunted that a nine-year-old spur might be a half-inch long. With such minute annual growth, imagine how bunched together new leaves can be in the spring. On a birch tree, two new alternate leaves at the end of a spur appear at first glance to be opposite. On a normal branch (not a spur), time will define leaf arrangement as the branch grows and separates neighboring alternate leaves. In high summer it is clear that birch leaves are neatly alternate in a plane since they spiral by 180° increments. But on slow-growing spurs, leaves may continue to appear opposite because the dwarf twig cannot adequately separate them.

Botany Booklet Chapter 5 – Simple or Compound Leaves

Although I chose not to begin the botany booklet with this chapter, I recommend this as your first anatomical consideration when studying a plant in the wild: Is the plant simple or compound? All of the leaves we have drawn or studied up to this point are *simple*, meaning each leaf is comprised of one single unit or blade. A *compound* leaf is composed of multiple blades called *leaflets*. How does one know if a blade is a leaf or a leaflet?

Stepping back from a plant to look at its entirety can give a clue. Does the plant appear to group its blades into clusters of a repetitive number? Like poison ivy or clover – both of which usually group in threes. Or are there two consistent numbers? For example, do you see blade groupings in fives … and some sevens … and maybe even nines? This would be common with hickory. Such repetition will catch your eye, and it suggests a plant with compound leaves.

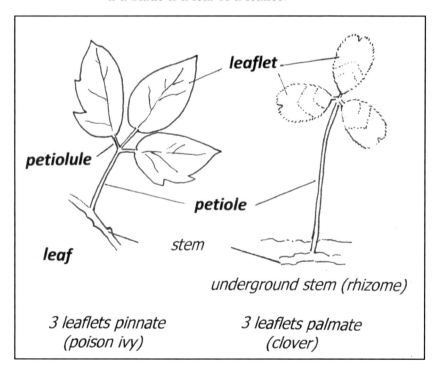

3 leaflets pinnate (poison ivy) 3 leaflets palmate (clover)

Some leaves have such deep sinuses that they might be evolving into compound leaves, in which case a lobe may one day become a leaflet. Because of the fine line between leaves divided into leaflets and leaves divided into deeply cut lobes, some books lump both types together into a common category called "***divided***."

There are two types of compound leaves, labeled according to the arrangement of their leaflets. If a compound leaf is somewhat elongated, having its leaflets attached like the filaments of a feather to its quill, that compound leaf is *pinnate* – after the Latin word "*pinna*" for "feather." This compound leaf might at first resemble a branch with opposite (or almost opposite) leaves, but one feature negates that supposition: There is a blade at the end of the so-called "branch" and it is alone without a mate. That **terminal leaflet** is a clue that you are looking at a **pinnate compound leaf**.

But keep in mind that it is possible for pinnate compound leaves not to possess terminal leaflets, which might have been eaten by animals, or might have fallen off (black walnut trees tend to do this, as shown in the photo section), or, as with the vetches, might be replaced by a **tendril** curling from the end like a living tentacle in search of a support around which to twine.

There is no distinction as to leaflets being opposite or alternate. You'll often find both cases on the same pinnate compound leaf. What the leaflets attach to (the extension of the petiole) is called the ***rachis***, which is analogous to the ***mid-rib vein*** of a simple leaf. (You will learn about the mid-rib vein in Chapter 6 of the Botany Booklet.)

The other type of compound leaf is ***palmate***, named for its similarity to the human hand. Our fingers radiate from a common spot: the palm. A palmate leaf has leaflets that also connect to a common spot, like the spokes of a rimless wheel, all meeting at the hub. There is no terminal leaflet because the design is radial. When determining if a plant with compound leaves has opposite, alternate, or whorled leaf arrangement, remember to consider the leaves … not the leaflets.

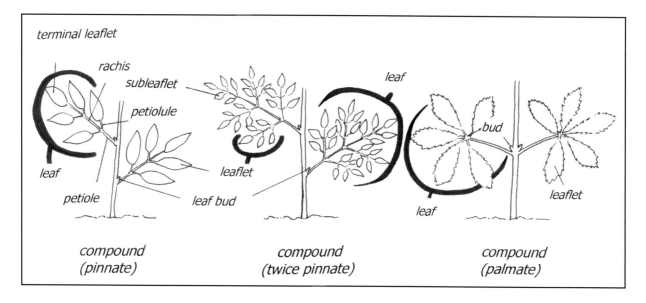

compound
(pinnate)

compound
(twice pinnate)

compound
(palmate)

But what about a plant with 3 leaflets? Is it pinnate or palmate? Look back at the pictures of poison ivy and clover. Poison ivy is pinnate. Its terminal leaflet stalk (petiolule) is longer than the other leaflet stalks and all three leaflets are somewhat pointing away from the leaf stalk. Clover is palmate, its leaflets more evenly spaced around the 360 degrees of a circle.

One factor tends to confuse students: "Is that a palmate compound leaf I'm looking at, or is it a whorl of leaves attached to a stem?" Generally, the best way to differentiate is to remember that whorls usually come in layers, like the stories of a building. If it is a stem with a whorl of leaves, the stem will likely continue growing up to the next node where another whorl of leaves will form. A compound palmate leaf's leaflets are structured like the ribs of an umbrella. That leaf will not add on double- and triple-decker redundant umbrellas. (One exception to this is trillium, with its single-tiered whorl. The helpful clue that defines trillium as a whorl of leaves is the flower that grows above the whorl. Leaves don't normally sprout flowers, but stems and branches do. Thus, trillium makes a single whorl of three leaves with a flower topping the stem.)

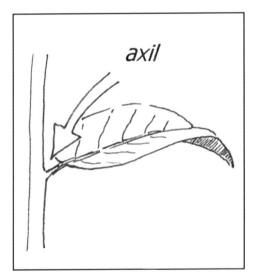

axil

If you are in doubt as to whether a tree or shrub is simple or compound, here's another point to check. When leaves open in the spring, next spring's leaf buds are already set and visible on most trees. These buds usually sit just above the present petiole attachment to the branch or stem – in the **axil**.

Check for a bud near the base of what you suspect is the compound leaf's petiole. If it's there, the whole complex structure below it is a compound leaf with leaflets. To double-check your assessment, inspect the angle where the leaflet stalk (petiolule) meets the rachis. If, indeed, this is a leaflet of a compound leaf, there will be no bud at this juncture. Leaflet buds do not exist.

If you now have a good grasp of the compound leaf, let's take it a step further. A plant can be twice compound … or thrice. A twice-compound leaf has its leaflets divided into sub-leaflets.

 ## Speaking the Language of Plants - This is a good time to close the book and go outside for a plant walk. (If it's winter, you'll be limited to the evergreens.) Stop at every herb, tree, weed, wildflower, and shrub to determine first if it is simple or compound. Then make decisions about its leaf shape, margin, attachment, and arrangement by referring to your booklet. You may begin to feel that you have learned a new language … the language of plants. And because you now speak this language, you will see plants more clearly, simply because you know what to look for.

Botany Booklet Chapter 6 – Veins

Primitive mosses have no vascular system, which is why they feel sponge-like (to retain water in the exterior maze of their fluffy structure). This fact more often than not gives credence to the adage about moss serving as a direction indicator. Moss generally covers more of the low, north (darker, moister) side of tree trunks, where sunlight (especially in winter) cannot dry out the plant. When, in the course of natural history, plants developed internal conduits for water dispensation, some plants were able to inhabit dryer sites, but mosses still follow their old habits.

Reading Direction from Mosses
 – Stuff a compass into a pocket, walk into an open well-sun-lit forest, and study the bases of ten different trees spread out over a half-acre. Note which side of each tree has the most prolific growth of moss, both in concentration and height on the trunk. In my area of hickory-oak-maple forests, the majority of those trees adhere to the old adage that moss prefers the north side of a tree. The minority that demonstrate the exception to the rule are affected by shadows from neighboring trees. Check your findings with a compass to see if the adage is helpful in your area. Take students to a different part of the forest and have them read the mosses there to re-determine the four directions.

These water tubes that we call "veins" start at the roots of vascular plants and travel upward to provide water and dissolved minerals and nutrients to the leaves. There, in conjunction with gases sucked in through pores and with absorbed sunlight, food is manufactured. In trees these up-conduits (the **xylem** tissue) comprise the outermost woody part of the trunk, the **sapwood**.

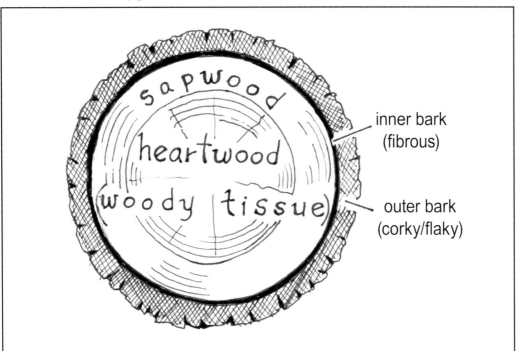

Down-conduits distribute the leaf-made food (as sap) to all parts of the plant, including the roots. This network of tubes is the **phloem**. Together with the green **cambium layer(s)** (which produces new conduit and new sapwood tissue) these down-tubes comprise the **inner bark** of a tree. It lies just beneath the outer bark.

You will learn later in this book the invaluable contributions that inner bark (or bast) offers to practitioners of primitive skills. Inner bark of certain plants can be made into cordage. Inner bark can also be a source for medicines. Some species of trees are said to produce edible inner bark, but in most cases the practical part to be eaten is the slushy cambium in spring.

These all-important "water pipes" (up-conduits of sapwood) and "sap canals" (down-conduits of inner bark) are the circulatory system of a tree. Due to a tree's "skin" of outer bark, inner bark tubes are not visible to us, but being so close to the surface of the tree these tubes are somewhat vulnerable. Humans' careless use with bulldozers, hatchets, nails, ropes, chains, and fence wire can be harmful and even devastating to a living tree. For example, when an uninformed novice camper whacks firewood against a living tree trunk for the purpose of breaking the wood into smaller pieces, he can crush those tubes and kill that side (rank) of the tree.

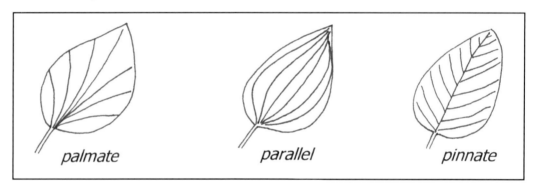

palmate   *parallel*   *pinnate*

These hidden conduits finally do show themselves where they fan out into the leaf in a web-like pattern we call "**venation**." Veins show up in three basic designs and in combinations of those three. All veins, of course, must enter a leaf blade from the same entrance point at the base (usually the petiole). If they flare outward in the blade, run side by side, then converge again at the leaf tip, they are called "**parallel**" veins. (They are not mathematically parallel … they curve.) Smaller transverse veins might connect these prominent ones making a more complicated picture, but the dominant veins define the pattern as "parallel." This is one of Nature's earliest patterns of venation; therefore, plants demonstrating it are very old, as a species.

sweetgum

Palmate veins fan out from the petiole but they do not converge again. If a palmate-veined leaf is lobed, prominent veins will run out to the tips of the lobes. Within a given lobe the venation might not be palmate. But, again, the leaf's vein pattern is determined by the prominent veins of the whole, not by the design seen inside a lobe.

You are most apt to misname a venation when the leaf is compound. Buckeye leaves, for example, are palmate-veined when you remember to consider the whole leaf, not the leaflet. (Each buckeye leaflet shows pinnate veins.)

The historically more recent and more highly evolved vein pattern is *pinnate*, shaped like a feather. The central vein that runs from petiole to leaf tip is called the ***mid-rib vein***. The question usually comes up from budding botanists regarding the opposite or alternate arrangement of lateral veins that branch from the mid-rib. Like leaflets, veins are never categorized in that way, as they can exhibit both opposite and alternate arrangements on the same blade.

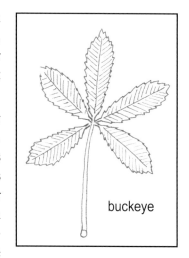

buckeye

Botany Booklet Chapter 7 – Cross-section of the stem

It is not necessary to damage a stem in order to determine its cross-section. Your eyes and fingers can make the assessment. If you see four corners connecting four even sides on the stem, it should be clear that you are looking at a ***square cross-section***. This might mean that the plant is a mint, since *all* mints have square stems; however, not all square-stemmed plants are mints.

If ever you do have need to actually cut across a stem to see its cross-section, use a *very* sharp knife on a firm workbench. A dull knife pressing into a stem can crush the tissue and distort your findings. Most plants contain a *pith* – a water storage compartment at the center of the stem. Piths can be relatively hard, soft, or sometimes hollow. When probing for a pith inside a tree branch, look for it to be darker or lighter than surrounding wood.

Other cross-section shapes of stems (or petioles, in the case of basal-leafed plants) are shown here. In the wild you will find additional cross-sections, but this list provides a good introduction to the possibilities. The cross-section is simply another observation tool as you accrue plant familiarity. One day in the future you may be in need of a fire-kit on a rainy day. No matter what technique you plan to employ for creating fire, the most challenging part of that kit during inclement weather is dry tinder. Pith material from dead plants might be a life-saver.

Every detail you can learn about a plant furthers your intimacy with that plant. Intimacy makes for indelible knowledge. Knowledge allows use. Use reinforces intimacy. It is a gratifying circle.

stem cross-sections

round

oval

triangular

square

single-groove

multi-grooved

winged

C-shaped

Botany Booklet Chapter 8 – Hairs

Leaves can be smooth (**glabrous**) or finely hairy (**pubescent**) or seriously hairy (**hispid**). There are even names for the angles at which hairs grow from leaves on specific plants. We'll simply concentrate on the *presence* or *absence* of hairs and their *locations* on the plant. Some plants have hairs only on one side of the leaf, some with hairs following the veins, others on the petioles or in the axils, and others at specific points along the stem.

Hairs can be a crucial factor in identifying a plant. I have had many beginner students confuse mountain laurel (*Kalmia latifolia*) – a shrub containing toxins lethal to humans – with sweetleaf (*Symplocos tinctoria*), a shrub also known as "horse sugar," which has delicious juices in its leaves. The presence of downy hairs on sweetleaf makes the differentiation easy. This same factor becomes important in distinguishing young milkweed (*Asclepias syriaca*, edible, hairy) from young dogbane (*Apocynum androsaemifolium*, dangerously toxic, hairless).

 <u>**Proof of Sugar-Manufacturing in Plants**</u> – Since we have not yet engaged in field identification of plants, find a plant expert in your neighborhood who can introduce you to sweetleaf (*Symplocos tinctoria*). Pluck off a leaf and note the downy hairs on the leaf's underside. Because these hairs cling stubbornly to the lining of the throat when ingested, fold the leaf several times to hide its hairy side. Chew the leaf only a dozen times to extract the Granny Smith apple flavor before spitting out the hair-infested remains.

sweetleaf (Symplocos tinctoria)

Botany Booklet Chapter 9 – Color

Plants synthesize impressive pigments – even in their non-flowering parts. The striking colors of fall leaves have actually been present throughout the growing season underneath all that green. When chlorophyll leaches out in the fall, a palette of hidden colors is revealed. At this point leaves have become nonfunctional to the deciduous tree, but they may go out in a blaze of glory.

Look closely at a plant specimen before the autumn change. Most plants show a variety of colors, even in their prime. You simply need to identify their shades and locations so that you become adept at recognizing that plant again in another geographic locale. The different shades of green may appear subtle in the early days of your plant education, but in time your eye will learn to distinguish between the multitudes of greens that tint the wild.

cranefly orchis

You might be surprised to discover a brilliant purple or silver on the underside of a leaf. (As an example, look in winter for Cranefly Orchis (*Tipularia discolor*) growing on a shady forest floor.) Such pigments serve as a reflecting shield to give a leaf a double dose of every light ray that passes through it. Or perhaps you will find a plant with outstanding pigments of pink or red or yellow at the nodes, or at the base of the stem, on the petiole, in the veins, or any number of places on the plant. Such colorations are identifying marks.

Botany Booklet Chapter 10 – Smell

Olfactory exploration can be one of the most telling probes into the world of Nature. Plant aromas can be both teasing and powerful in a teaching situation. In the brain, the olfactory center and memory vaults are positioned next to one another, which explains why long-forgotten, re-encountered smells can evoke lucid memories. But it is amusing how some of the most obvious smells elude us when they waft from the seemingly wrong package. Pine tar shampoo smell from the leaf of perilla (*Perilla frutescens*)… *Band-Aid* smell from a giant millipede (*Spirobolida*) … the varying smells of poisons exuded by daddy-long-legs (harvestmen "spiders"): alcohol, anti-freeze, banana, butterscotch, chocolate. We are so visually oriented that an incongruous source of a smell can confound our ability to identify its aroma.

The leaf of the pawpaw tree (*Asimina triloba*) smells just like a bell pepper (or like a tomato vine), but few students can put a name to the aroma. When given the answer to the riddle, students smile and then shake their heads at not recognizing the familiar scent.

The mature seed of black snakeroot (*Sanicula marilandica*) smells like an orange popsicle; a sassafras (*Sassafras albidum*) leaf like *Froot Loops* cereal; a spicebush (*Lindera benzoin*) leaf like grapefruit rind. The scent of some hickory leaves evokes the aroma of *Listerine* mouthwash.

On countless occasions I have watched a surprise scent from a plant win over a young student who then explores everything in her path with her nose. (**Draw the line at sniffing animal scat, because some droppings contain parasites that can be inhaled!**) This student discovers a tool she has always had, but now she

employs it to further probe the mysteries of Nature. Like a muscle that is exercised, her sense of smell will improve.

I believe it sets a bad example to traipse through the forest snapping off leaves thoughtlessly, crushing greenery to smell it, then discarding it like so much trash. I'll share with you my protocol for considerate leaf smelling: 1.) First, get on the plant's level. If it's an herb this means stretching out on your belly on the forest floor. Smell the leaf by approaching it with your nose. If no results, 2.) gently rub the leaf between thumb and finger, then sniff again. If no results, 3.) choose a leaf with prior damage, if available, and make a small tear in the blade without disturbing the rest of the plant. If no results, 4.) complete the tear, ball up the torn part, and crush it until moist. Smell. If still no results, 5.) open up the crushed leaf and wait a full minute while oxygen mixes with the released juices. You might get a surprise. If you get no results after this, you have exhausted the possibilities for field exploration in leaf aroma. For roots or twigs a scratch and sniff will suffice.

This "formula" for smelling plants is a rich lesson for children and adults alike. It reminds us not to be blindly destructive. It teaches us that we should not consider ourselves in control of all things at all times – especially the life and death of other organisms. It grants us an entry into the forest as allies rather than as dominators, and this elevates us to better relationships with Nature.

I have observed over the years that whenever I ask a group of students to sit on the earth, soon a majority of hands begins to unconsciously tear things apart – plants, sticks, leaf litter. Is it our nature to disassemble? Conversely, I have watched my veteran botany students keep their hands idle. These latter students seem more attentive to life and to the present tense. Perhaps they have learned to inhabit the forest rather than trespass into it.

You will find that some plants hold different smells in different parts of their anatomy. Sassafras roots, for example, do not contain the *Froot Loops* aroma. They smell like root beer.

You need to know which plants *not* to handle: mayapple, poison ivy, cactus, and stinging nettle, for example. Chapter 10 contains a long list of plants known to irritate human skin. You'll learn these as you come to know more about plants. Most plant books contain warnings. A good practice for your own safety is to wash your hands in a creek after each olfactory adventure. This gives your fingers a clean slate on the occasion of your next probe into another plant's scent.

One leaf that forces you to go all the way to the final smelling step is a mature blade of a black cherry tree, *Prunus serotina*. In summer the experience can be rich, like opening a jar of Maraschino cherries to an almond extract aroma. But don't let your nose be your eating guide. In the wild this aroma indicates cyanide (a poison) or one of its relatives. Livestock can die from eating the wilted leaves of storm-torn branches of black cherry trees.

Improving your sense of smell will come in handy later when you delve into the skill of tracking. Start getting "olfactory-strong" now. Let your nose go feral.

On the Trail of Plant Scents – Make a thorough examination of leaf

smells in a given area, say forty paces square, smaller if the woods are thick and larger if somewhat barren. Be careful not to handle those plants that could be harmful to touch. If you feel incapable of making such an assessment, it's time to call up a local plant enthusiast and ask if she can accommodate your beginning efforts by doing an identification walk with you. This is worth paying for or bartering. Have that person emphasize plants that are not touch-friendly.

When the olfactory exploration is complete, you and your child (or your students) each commit to the names of three leaves that you can confidently identify by smell, even if this means making up a name for each plant. In teams of two, one person lays out five different crushed leaves for the other to smell while blindfolded. The three named leaves are among these. Identify the three "learned" aromas. Then reciprocate by having teammates switch roles.

Botany Booklet Chapter 11 – Flowers

Almost all the layperson books on herb identification rely on the presence of a flower for the reader to go through the steps to successfully name the plant. In reality, more often than not, when you want to identify a plant, there is no flower present. You might resort to thumbing through a book, hoping to spot the correct drawing or photo.

Every serious botany student eventually owns a well-used field guide for plant identification. Unfortunately, using one of these books successfully depends upon the presence of a flower. For all my students I recommend *Newcomb's Wildflower Guide* by Lawrence Newcomb, published by Little, Brown and Company, because his book is user-friendly to the beginner and contains a "key" (a systematic series of questions that leads to identification) that allows you to make assumptions about a plant's flower if the flower is not present. For example, you might see the remnant of a seedpod divided by four seams – or perhaps four dried, woody parts radiating from beneath a seed receptacle. From this information, it would be reasonable to assume that the flower had four petals.

We shall consider a few basic flower designs. Flowers that show **radial symmetry** are wheel-shaped with a central hub and any number of petals extending from that hub. These are often referred to as "**regular**" flowers. An insect standing at the center of a regular flower (like an aster or anemone), no matter which direction it faces, sees virtually the same flower-scape.

If you try to examine a sunburst of petals around a flower's center, you might not be looking at

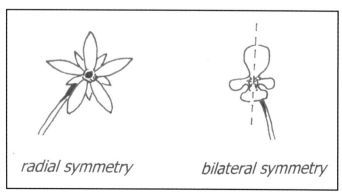

radial symmetry   bilateral symmetry

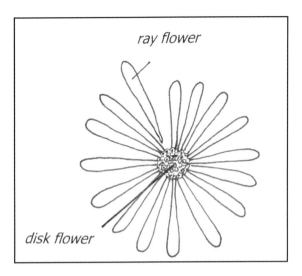

ray flower

disk flower

petals at all. Plants belonging to the composite family (like sunflowers and asters) produce conglomerates of many flowers that are arranged together into a typical single flower shape. With a daisy, for example, one of its many white "petals" is not really a petal but a complete flower called a **ray flower**. If you were to gently slip it free from its neighbors you would see its reproductive organs attached to the narrowed base. The yellow button-like center is a mosaic of tiny **disk flowers** arranged in a tight spiral. What you thought was a single flower might be more than a hundred disk and ray flowers all put together in the guise of a single flower.

On some blossoms – like flowering dogwood – what appear to be "petals" might actually be modified leaves called "**bracts**." Bracts are usually situated at the base of a flower or flower cluster, but in the case of dogwood there are four large white bracts. The true flower is the smaller green structure located at the confluence of the bracts.

Bilateral symmetry describes those flowers that have a vertical axis dividing mirror images (like the human body, whose axis of symmetry runs from the crown of the head, down between the eyes, through the navel and beyond). A bilateral flower is often referred to as "*irregular*." An insect at the center of this flower will see similar mirror-image scenes in only two directions.

The petals of a flower are even more "attractive" than they might appear to us. Petals are decorated with colors whose wavelengths lie outside the visual range of humans. These hidden colors guide flying insects directly to the business center of a flower where reproductive parts await pollen couriers. These colored spots serve like landing lights on a runway to orient a pilot.

At the center of the collective petals (*corolla*) of a typical radial flower stands a tower waiting to receive pollen. This female receptacle is the **pistil** and houses the ovary. Its top opening is the **stigma**, which sits atop the tube-like **style**. Standing around it are

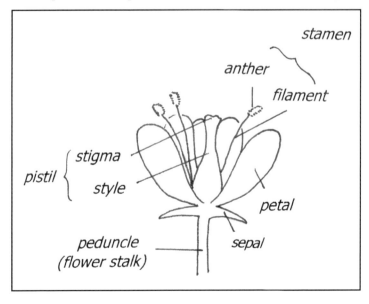

stamen

anther

filament

stigma

pistil

style

petal

peduncle
(flower stalk)

sepal

male **stamens**, each composed of a stalk called a "**filament**" and the pollen-producing organ at its tip, the **anther**. As you can see, this single flower example is both male and female, but some species of plants make male flowers and separate female flowers, sometimes on the same plant and sometimes on separate plants. (A single persimmon tree, for example, has a gender … it makes either male or female flowers, but not both.) Beneath the petals of a flower you might find green **sepals**, collectively, called the "**calyx**."

The arrangements of flowers on a plant add more unique words to our vocabulary. The drawings below illustrate the patterns you will commonly encounter.

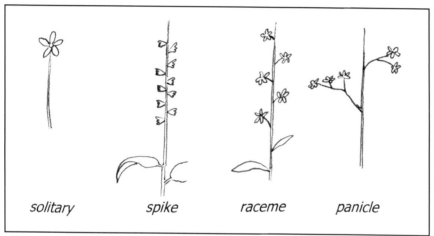

solitary   spike   raceme   panicle

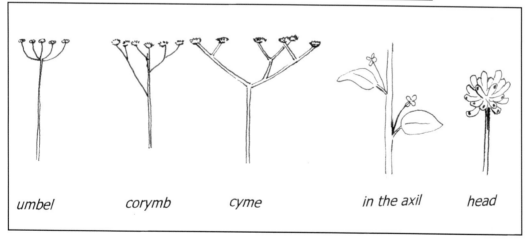

umbel   corymb   cyme   in the axil   head

This brings us to the end of the self-made Botany Booklet. But you will have more information to add to its pages, both from the field and from reference books. Add to your personal book the new terms you learn to use, and make lots of drawings. And now for the cover illustration …

An Educated Drawing – Remember that first drawing on the inside of the

front cover of your Botany Booklet? *Do not* now look at it. Return to the same plant specimen and draw it again, this time on the cover and now with a trained eye. Use each booklet topic as a guide in your artwork. When you are done, compare the "before" and "after" drawings to see proof of your botanical education.

Drawing the Treasure for a Hunt – Have each student choose a living leaf to draw outside. No two students should select the same plant species. Use the booklet chapters as a drawing guide, and render each portrait *in situ*. A step-by-step format is offered below. With the drawings completed, allow students to trade sketches and, in a new site, search for a plant of the drawn species. Depend upon the artist to assess the hunter's accuracy.

Making an Accurate Botanical Drawing

1. Measure the **length of the petiole**. Use a stick or finger-width as a ruler – whatever works with that particular leaf stalk. For example, if the petiole is as long as your pinky is wide, let's call that measurement "P" for "pinky." If you want to draw the picture larger than life-size, multiply P by the desired integer; say, 3.

2. Measure the **length of the leaf specimen** in terms of P. If the leaf is 6 times as long as the petiole, that measurement is 6P. In the drawing this becomes $3 \times 6P = 18P$.

3. Measure the **width of the leaf** in terms of P. If the leaf is 1/3 as wide as long, the width can be measured on paper by $3 \times 2P = 6P$. Be sure to draw the widest part of the leaf in the proper place along the length of the leaf.

4. Note how the **base of the leaf** flairs from the petiole. In the drawing, draw the proper angle of the flair of leaf blade at the base.

5. Note how the **sides of the leaf** meet at the tip. In the drawing, draw the proper angle for this convergence.

6. Note where **lobes** (if present) stick out relative to leaf length. Draw them at their appropriate locations.

7. Note the exact **outline of the sinuses**. Draw the proper length of sinus cutting into the blade and be sure to render the innermost part of the sinus accurately. Is it a U-shape, V-shape, teardrop shape?

8. Where do the **veins** run? Do they run out to teeth? Do they fork? Do they bend away from the margin of the leaf?

9. How many **teeth** (if present) lie between neighboring veins?

10. Draw one more leaf (outline only) to establish leaf **arrangement** on the plant: opposite, alternate, whorled, or basal.

Keying Out Plants – To begin using a field guide (book) to identify plants, start with *a plant that you already know*. Using the book's key, work your way to a page number in the book, where a drawing or photograph of your specimen is waiting for you. If you don't find it there, check neighboring pages. If this too fails, you have the luxury of turning to the index to find the page number of your known plant. Flip to that page and read its description to see where you went wrong in your assessment of the plant's anatomy.

<u>Keying out 5 unknown plants using a Field Guide</u> – Go for a
summer walk in a field where you are likely to find wildflowers. Sit down next to the first unknown wildflower you encounter. Going through the "key" of the book, work your way to the page where your unknown plant is drawn and described. In a new notebook dedicated to your life-study of plants, make a drawing of the plant and note its location. Repeat four times with four different plants. Later, when you learn about foods and medicines, you'll have practical information to write on these pages. Perhaps you will return one day to make use of one of these plants.

<u>Comparing the Standing People to Ourselves</u> – Besides veins,
what other anatomical similarities do we share with the Standing People? Gather your class in the shade of a fine tree specimen. Examine the roots. There are several reasons that they spread out – one of those being for support … like our plantigrade feet. Beneath the soil the roots divide and subdivide and the smallest rootlets are covered with root hairs that produce acids to dissolve nutrients present in the soil. Once dissolved, these nutrients are then available to be absorbed into the root so that they can be delivered to the leaves where food is produced. What part of our bodies works like those root hairs to break down compounds? Make a human body poster and a tree poster, and connect their analogous parts by pinning lengths of yarn from one poster to the other. Here are some parts of the tree to discuss, to research, and to ponder how their functions compare to the workings of our own bodies:

1. Root hairs (acid production, dissolving, absorbing) = tongue, teeth, lips, stomach, small intestines.

2. Lateral roots (structural stability for uprightness) = feet.

3. Roots in general as graspers (clutching soil or boulder) = hands and fingers.

4. Sapwood (swallowing water and dissolved nutrients) = esophagus.

5. Wood in the trunk and branches of the tree (structural rigidity for uprightness) = bones.

6. Pith at the core of the wood (water storage, flexibility) = bone marrow.

7. Inner bark (distribution of food) = blood vessels.

8. Outer bark (protection) – skin, calluses, hair, nails.

9. Open wounds in the bark (injury) = cuts and abrasions (consider disease and infection potentials).

10. Unusual bark formations sealing old wounds = scabs, scars.

11. Swellings and lumpy growths where insects have invaded to lay eggs = tumors, cysts, cancer.

12. Branches (as extenders) = arms.

13. Leaves (the light/shade factor) = eyes tell us which way to go as we weave through a thick forest.

14. Leaf hairs (protection from insects and prolonged sun exposure) = eyelashes, eyebrows, scalp hair.

15. Leaf and bark pores (stomates and lenticels) – mouth, nose, bronchial tubes, lungs, skin pores.

16. Thorns = vocal tones, posturing, hostile or defensive facial expressions, finger nails, fists.

17. Flowers = cosmetics, hairstyles, cologne and perfume, courting language.

18. Fruits, nuts, seeds = embryo

"And the world cannot be discovered by a journey of miles, no matter how long, but only by a spiritual journey, a journey of one inch, very arduous and humbling and joyful, by which we arrive at the ground at our feet, and learn to be at home."

Copyright © 1991 by Wendell Berry, *from* The Unforeseen Wilderness.
Reprinted by permission of Counterpoint.

CHAPTER 3
Venturing Out into the Known and the Unknown

Almost everyone – no matter his or her level of outdoor experience – is able to identify on sight at least a few trees. Let's use this knowledge to our advantage in our quest to understand the identifying process. We are not yet ready to harvest plants for use as food, medicine, or craft; but a few practical previews are listed here to get you excited about what is to come.

A Closer Look at Some Old Friends Among the Standing People

– Gather your self-made Botany Books, paper, and pencils and go outside for a walk to look for trees *whose names you already know.* At each tree, make a close study of living leaves. If you cannot reach them, inspect a dead leaf found on the ground. Chapter by chapter in your booklet, consider which botanical terms apply to this leaf. First, are the leaves compound or simple? Once you have determined this, make notes about your observations. What is the shape of the leaf? How is the margin described? And so on. After you have answered the questions that each of your topics covers, compare your findings to the information in the list below, where I've included trees you will likely encounter. If you are unable to identify any trees, seek help from a tree-savvy friend.

 Oak (*Quercus spp.*) ["spp." designates the plural of "species" and using it here refers to different types of oaks] – With oaks, of which there are so many species, it is not so important at this point in your plant education that you determine which particular species you are studying. But do learn to differentiate the two

large groups – red vs. white oaks. Generally speaking, most oaks have lobed leaves, and the shape of these lobes gives us an old adage to help distinguish reds and whites: *White* oaks have rounded lobes like the *white* man's musket ball; *red* oaks have pointed lobes like the *red* man's arrowhead. (Reds also show bristles at the ends of lobes, but there are exceptions to this rule.) This information will become pertinent during your harvest of acorns for food in the next chapter.

Be warned: Oaks leaves can be deceptive. To compensate for a shady level in the forest, low-growing red oak leaves often expand their surface area by filling in their sinuses with leaf tissue. As a result lobes can become quite rounded. Usually, the defining factor is the presence or absence of the aforementioned bristles at the leaf or lobe tips. To determine either, always examine many leaves on the same tree.

To define leaf shape, draw an imaginary line around the leaf (ignoring the sinuses) letting the tips of the lobes serve as the dots you connect to make a drawing of a lobe-less leaf. This will help you see where the widest span of the leaf lies.

Leaf margins of the <u>lobed white oaks are smooth</u>. The chestnut oak of the mountains has <u>rounded teeth</u>. All oak leaves are <u>simple</u> and grow <u>alternately</u>. Petiole lengths vary from one species to another. Veins are <u>pinnate</u>. Many oak leaves show <u>hairs</u> when young, often on the underside of the blade, but these hairs may be lost as the leaf matures, unless tufts of hairs are retained in the <u>axils of the veins</u> on the underside of the leaf.

Generally, white oaks have flaky, light-gray bark that breaks and scatters when you rub your hand up the trunk. (Chestnut oak is an exception.) Red oaks have firmly attached dense bark.

A field guide to trees of your region can help you determine exactly which oak you are studying. To help you narrow down the possibilities, some of the more common oaks of Southern Appalachia are these: among the whites – white oak, chestnut oak, and post oak (edge of a field); among red oaks – scarlet, black,

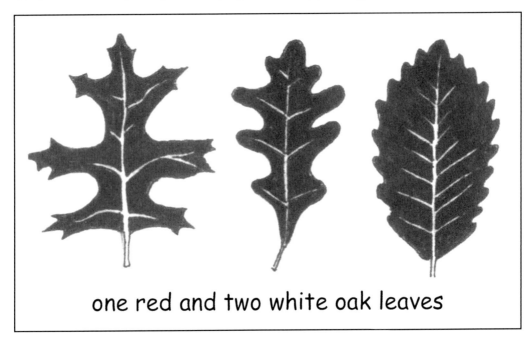

one red and two white oak leaves

southern red, northern red, blackjack, and water oaks. Further south on the coastal plain: live, turkey, laurel, sand, and bluejack oaks.

Tulip Magnolia or Tulip Tree (*Liriodendron tulipifera*) – Though most people incorrectly refer to this as a poplar ("yellow" or "tulip" poplar … even the U.S. Forest Service and many books have adopted these misnomers), you'll want to know this tree by its true name. This knowledge will benefit you when you delve into fire-making and edible vs. inedible inner bark. Contrary to popular belief, the tulip tree is not a poplar. It is a magnolia.

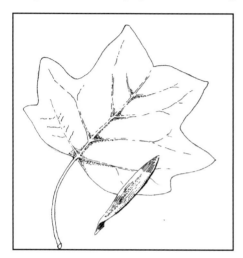

The large <u>simple</u> leaf with 4 to 6 <u>lobes</u> is <u>alternate</u> and resembles a headless wide-tailed bird in flight – the long <u>petiole</u> taking the place of the head. But the most eye-catching feature is the broadly <u>notched tip</u>, which makes this leaf unique among all other native trees (which have pointed or rounded tips). Veins are <u>pinnate</u>.

The gray bark is smooth on young trees but furrows with age. The trunks grow fast and straight. On the trunk bark, dark scars "hang" over branches (and former branches) like inverted V's seared into the trunk bark. The large tulip-shaped flowers are green, orange and yellow and develop into advanced scale-less cones composed of <u>winged seeds</u> that attach to a central <u>spine</u>. When found, this spine resembles a tiny wooden sword with a dimpled blade. (Each dimple marks where one seed articulated with the spine.)

In time you will learn to place a high value on the tulip tree. It is an easily found source of abundant fibers for tinder and cordage. Its spring bark can be removed in sheets for various crafts.

Pine (*Pinus spp.* and the Pine Family) – Besides the many so-called "pines," this large family of conifers includes firs, larches, spruces, and hemlock trees. Most people recognize a pine tree by its needle-leaves, but there are other trees with needle-like foliage: juniper and cypress (both often incorrectly called "cedar"), yew, and true cedar (which is not native to North America but possible to find here due to landscaping with exotic plants).

Many pine leaves are <u>compound</u> with needles bunched into small bark sheaths called "fascicles." Counting the needles in a bundle is part of the identification process. Most compound pines have either 2 or 3 needles bundled together. Some needles are straight and others twisted; some limp and some stiff. Each species has a characteristic length of needle. Needle bundles are <u>alternate</u>. Vein patterns of pine are never discussed in field guides, but let's name the pattern for what it is: <u>palmate</u> … because the leaflets (needles) do fan (if only slightly) from the fascicle.

In the Appalachians, **white pine** (*Pinus strobus*) is easily identified by its <u>bundles of five needles</u> and its orderly <u>whorls</u> of limbs (which give the tree a tiered structure). Outer branchlets can grow opposite or whorled. Needles are soft and

flexible 3" to 5" long and milky blue-green. The needle litter on the ground is luxuriant to bare feet. Cones are long (up to 8") and slender and generally appear only at the outer periphery of branches.

In the piedmont, **loblolly** pine (*Pinus taeda*) has dark flaky bark (like stacks of scorched potato chips) that can take on a reddish tint with age. Needles are <u>bundled in 3's</u> and 6"- to 9"- long, stiff and dark green. Mature cones have very sharp prickles.

Longleaf pine (*Pinus australis*) of the coastal plain also has <u>3 needles in a bundle</u> but the needles are 10" to 18" and crowded along the branch. Cones can be 10" long.

As you'll learn in the next chapter, all the native pines of the East offer a variety of foods, an excellent base for glue, aggressive kindling, cordage, and a drawing poultice for splinters or infection. While most pines are poor materials with which to create fire by friction, dead wood of white pine, hemlock, fir, and short-leaf pine make stellar kits.

Willow (*Salix spp.*) – Since most people are familiar with weeping willow (*Salix babylonica*, a Chinese import), it is not a stretch for them to recognize a native willow that grows by streams, ponds, and septic fields. Black willow (*Salix nigra*) looks similar to its "weeping" relative but lacks the long whip-like branches that droop to the ground. Leaves are <u>simple</u>, <u>lanceolate</u>, <u>alternate</u>, and finely <u>toothed</u>. Veins are <u>pinnate</u> and paler than blade tissue.

The trunk can be quite dark or gray with forking ridges of long strips that delaminate to a sometimes ragged

black willow (Salix nigra)

sweetgum

appearance, overlapping like sideways shingles. This tree manufactures salicylic acid, which is the basis of aspirin. The wood is prized for fire-creation.

Sweetgum (*Liquidamber styraciflua*) – Many people confuse maple with sweetgum, unless the latter's large spiked seed balls are underfoot. A sure way to differentiate between the two trees is to study the leaf and branch arrangement. Maple demonstrates <u>opposite</u>, sweetgum is <u>alternate</u>.

Leaves are <u>simple</u>, <u>star-shaped</u> with <u>palmate</u> veins (lobes are pinnately veined). The margin is neatly <u>toothed</u>. <u>Petioles</u> are long. Crushed leaves give off a <u>strong scent</u> somewhat reminiscent of crushed goldenrod.

The bark is soft and <u>corky</u>, easily dented by a thumbnail. You might find <u>wings</u> of flanged bark following the branches. And, of course, look for the "gumballs." Leaves and sap contain an antiseptic. The wood, if properly dried, is useful in a fire-kit.

<u>*First Venture Out into the Unknown of Your Yard*</u> – Now we

move to plants that you might have noticed but whose names you do not know. To help you get started in botanical field identification, here is a brief guide to seven plants you will have a good chance of encountering around your home. Refer back to the chapters in your booklet to review the anatomical parts described below and, without using any other reference, simply roam around your yard. Use the photos in this section to help you search. When you think you might have located one of these seven, make a drawing of the found specimen in your notebook. Then by referring to a field guide – book or human – verify your findings. Practical uses of these plants are mentioned here in passing only to give you a tantalizing taste of the groceries, medicines, or craft materials that grow so close to your home. Before actually utilizing these gifts, you need to invest more time in the practice of identifying. Be patient. We'll soon get to the practical uses of these plants.

In most lawns – except those violated by herbicides – you are likely to find these plants:

1.) **Broad-leaved plantain** (*Pantago major* or *Plantago rugelii*) has only <u>basal leaves</u>, which are <u>ovate</u>, haphazardly <u>toothed</u> or <u>smooth</u> and ranging in size from a small hen's egg to somewhat larger than your hand. These leaves often lie nearly flat on the ground. Their <u>parallel veins</u> are prominent. At the center of the rosette look for furled pale-green new leaves standing more upright. The <u>petiole</u> is shaped like a half-pipe "waterslide." At the base of the petiole you may find a purple-pink pigment, which makes this specimen **red-stemmed plantain** (*P. rugelii*). (Note: This plant would be more correctly named "red-stalked plantain.") If the pigment is

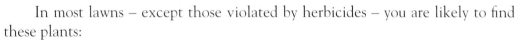

absent, you have found **common plantain** (*P. major*). The sure proof of identification of either type of broad-leaved plantain comes by *gently, patiently* pulling apart a mature leaf blade from its petiole and watching the tough little veins pull out of the leaf. The veins of young leaves may be too tender to demonstrate this trait, so experiment on older leaves. If the plant is in flower, it sends up a slender spike of tiny flowers displaying off-shades of white, green, or brown.

This is an Old World plant whose seeds inadvertently found their way to this continent. When you advance to wild foods and medicinals in the next chapter, you'll be happy to learn of the many uses of this plant, one of those being a remedy for a bee/wasp/ant sting.

broad-leaved plantain (Pantago major or Plantago rugelii)

Any plant whose name includes the word "plantain" was at some time in history *believed* to be useful as a poultice to draw a foreign substance or infection from the body, but we now know that this name assignment was not always appropriate. Broad-leaved plantain, however, works as an effective poultice to neutralize sting-venom. In addition, it is a food, insect repellent, intestinal cleanser, and anti-inflammatory. More on this later.

2.) **Red** (*Trifolium pretense*) and **white clover** (*T. repens*), due to their "lucky" four-leaflet anomalies, are known to most American children. Their <u>palmate, compound leaves</u> might appear to be <u>basal</u>, but look for a stem that creeps along the ground. Each alternate leaf is composed usually of three leaflets. Each leaflet is very <u>finely toothed</u> with a subtle <u>notch at the tip</u> and often with a pale green <u>chevron</u> spanning its width. Thin, tubular, <u>irregular flowers</u> of white to light pink or red are clustered in a <u>head</u> on a stalk slightly longer than a leaf stalk. Both clovers are considered edible after the proper preparation, but undetectable diseases or toxins make clovers problematic for consumption. Three plants often confused with these plants are hop clover (*Trifolium agrarium*), black medick

red clover (Trifolium pretense)

(*Medicago lululina*), and wood sorrel (*Oxalis montana*). These three have distinct above-ground stems and <u>yellow flowers</u>.

common dandelion (Taraxacum officinale)

3.) **Common Dandelion** (*Taraxacum officinale*) is another plant known to all children who have plucked its flower stalk – after it has gone to seed – to blow a fleet of parachute seeds into the wind. Dandelion has only <u>basal leaves</u>, which are extremely and irregularly jagged with <u>teeth</u> that tend to point back toward the leaf base in barb-like fashion. At the top of the <u>smooth hollow mature flower stalk</u> is a <u>head</u> of petal-like yellow ray flowers, each of which produces a parachute and seed. You can gently pull out a ray flower to see the premature parachute folded at its base. Just before the head goes to seed, the stalk engages in a second growth-spurt to better catch the wind currents above surrounding vegetation. Dandelion boasts dyes, medicinal uses, and, as we shall see, edible parts with high nutrient value.

4.) **Common Chickweed** (*Stellaria media*) is a low weed (under 12 inches, usually) that can grow upright but tends to recline, so that a colony looks like it would make a nice mattress for a rabbit. Its <u>opposite, ovate leaves</u> are <u>smooth</u> and have <u>long petioles</u> on the lower stem but <u>short petioles</u> on the upper stem. The stem is <u>succulent</u> with rows of <u>hairs</u>. <u>White starry flowers</u> grow singly

common chickweed (Stellaria media)

or clustered at the top of the plant. The 5 <u>petals</u> are shorter than the <u>sepals</u> and are so deeply cleft they appear as 10. New growth is edible raw or cooked. [One species, mouse-ear chickweed (*Cerastium vulgatum*) – so named for its very hairy leaves – requires cooking.] Chickweed is so nutritious that it has earned a reputation as an appetite suppressant for dieters.

Second Venture Out into a Sunny Vacant Lot or Barnyard

1. **Pokeweed** (*Phytolacca americana*) emerges in spring with its leaves pointing skyward. These large <u>ovate leaves</u> show a vibrant lime green hue and attach to a <u>succulent</u> stem in <u>alternate</u> fashion. A toxic <u>red-purple pigment</u> climbs the stem as the plant matures. <u>Racemes</u> of <u>white to pinkish</u> flowers (with <u>five sepals</u> that look like petals) develop into long clusters of dark purple berries, and the <u>branching stem</u> eventually assumes a beautiful rich magenta and is easily broken due to its <u>fragile pith</u>. Young spring leaves are delicious as a cooked green once toxins are dispelled. **Do not dine on this plant before referring to Chapter 6!**

pokeweed (Phytolacca americana)

evening primrose (Oenothera biennis)

2. **Evening primrose** (*Oenothera biennis*) is most easily identified at the end of its second year when its dried fruit capsules remain open on the <u>branches</u> of the stem. These capsules resemble one-inch-long, partially-peeled, wooden bananas. Primrose is a <u>biennial</u> – living two years. In its first year it makes a <u>basal rosette</u> of <u>lanceolate</u> (or long elliptical) leaves with a <u>pink pigment</u> bleeding into the mid-rib vein from the petiole. It takes experience to identify the first-year plant, whose taproot and young leaves can be prepared as a food.

In the second year a tall stem rises to advertise the bright <u>yellow flowers</u>, which have <u>four large petals</u> and a cross-shaped <u>stigma</u>. <u>Sepals</u> point backward away from the flower. Stem leaves are <u>alternate</u>. The seeds and petals are also edible. The dried stem is an excellent hand drill for fire.

curly dock (Rumex crispus)

3. Curly Dock (*Rumex crispus*) has robust, dark-green, textured leaves irregularly <u>toothed</u> on <u>wavy margins</u>. These leaves grow <u>alternately</u> on a <u>multi-grooved stem</u>. Look for purple-pink pigment in the midrib vein. The small dangling <u>green flowers</u> are arranged in <u>whorls</u> on <u>racemes</u>. The seeds have three small <u>heart-shaped wings</u>. A crushed leaf can be rubbed into the skin after fire ant or nettle stings to immediately neutralize the acid and end discomfort. Other uses include a dye, styptic, anti-fungal, and a variety of foods.

4. Common Burdock (*Arctium minus*) is a large coarse weed that produces robust, <u>ovate, alternate leaves</u> – the lower ones usually <u>heart-shaped</u> and having <u>hollow petioles</u>. The margins can be toothed or smooth. A biennial, it produces in its second year <u>violet-purple flowers</u> in <u>bristly heads</u> that arise from the <u>axils</u> and top of the stem. Once gone to seed, the

common burdock (Arctium minus)

fruit is a bur that readily sticks to clothing with its hooked spines. As will be discussed later, foods can be prepared from the root, young leaves, basal leaf stalks and flower stalks. (**Rhubarb, whose leaves are poisonous, could easily be mistaken for burdock before the flowering stage!**)

5. There are quite a few plants called "**bedstraw**," including a robust species (*Galium aparine*) called **cleavers** or **goose grass**. Its <u>square stem</u> and leaves are prickly with tiny spines that can stick to clothing so that the plant can be inadvertently torn from the ground when trod past. The <u>oblanceolate leaves</u> can grow to 3" and are arranged in <u>whorls</u> (4 to 8 leaves). <u>White 4-petaled flowers</u> arise from the <u>axils</u>. Cleavers is a source of food, sunburn treatment, and "coffee."

bedstraw (Galium aparine)

thistle (Cirsium spp.)

6. **Thistles** (*Cirsium spp.*) have <u>alternate</u> leaves that are extremely spiny – elevating the term "serrate" to a new level. Handling these plants can draw blood. Most are biennials that begin life as a basal rosette and produce a stem in the second year. The majority of thistle flowers are purple and develop as large "tufts" inside bristly "vases." It is a most distinctive plant that can grow taller than a man. Leaves, stem, stalks, and root can all be prepared as food.

sumac, smooth (Rhus glabra)

sumac, winged (Rhus copallina)

7. **Sumacs** (*Rhus spp.*) are shrubs with <u>pinnate</u>, <u>compound</u>, <u>alternate</u> leaves. Of the several plants named "sumac," only one is taboo to touch. **Poison sumac** (*Toxicodendron vernix*) **grows in low wet woods of the coastal plain and in swamps or bogs and on higher ground where water seeps from rocks. Its leaflets have**

smooth margins. Leaflets of **smooth sumac** (*R. glabra*) (here "smooth" refers not to leaflet margin but to a <u>lack of hairs</u> on the branches and petiole) and **staghorn sumac** (*R. typhina*) with <u>hairy branches</u>, <u>petiole</u> and <u>rachis</u> are <u>toothed</u>. Another safe species, **winged sumac** (*R. copallina*) has mostly smooth margins on leaflets but shows <u>"wings" of leaf material</u> on the rachis between leaflets, distinguishing it from the toxic species. The <u>yellow-green flowers</u> of the safe-to-handle sumacs grow at the <u>ends of branches</u>. These safe berries turn <u>deep red</u> in erect, candle-flame-shaped clusters from which a delicious drink (the Cherokees called it "qualla") can be made. Poison sumac berries dangle from leaf axils, are grayish-white, and do not turn red.

Third Venture Out Along a Creek in a Piedmont or Montane Forest -

1. **Yellowroot** (*Xanthorrhiza simplicissima*) grows on well-drained stream banks and often leans out over the water. Its <u>pinnate compound leaves</u> resemble those of celery with <u>irregularly toothed</u> margins that appear snipped by scissors to a rough fringe. The leaves are <u>alternate</u> but often are bunched tightly at the top of a gray, knobby, wooden stem, where they can fan out like the ribs of an umbrella. A fingernail scrape into the bark of a root reveals a bright <u>yellow pigment</u>. The small, <u>purplish, star-shaped flowers</u> droop in <u>racemes</u>. Chemicals in the root are effective in soothing irritated mucosa, such as the cases of mouth sores and stomachache.

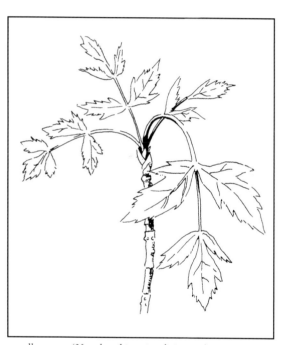

yellowroot (Xanthorrhiza simplicissima)

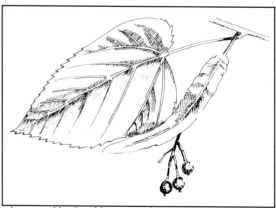

basswood leaf and berries with wing (Tilia americana)

2. **Basswood** trees (*Tilia americana*) are seldom found far from water in the South. Their large <u>heart-shaped</u> leaves have an asymmetrical base — one lobe of the "heart" being larger than the other. Leaves are <u>alternate</u> and <u>serrate</u>. The bark is similar to the tulip magnolia. Its cluster of small, <u>five-petaled, creamy-white to yellowish-white flowers</u> produces dark berries, which are guided to earth by a <u>leafy wing</u> that twirls and

catches the wind. Basswood offers food, medicine, cordage, tinder, carving wood, shelter shingling, basketry, and a fire-kit. **Do not confuse with mulberry trees, which contain a milky, latex-rich sap.**

3. **Jewelweed** (*Impatiens capensis*) prefers a place with open light by water. The very <u>succulent</u>, pale-green stem shows <u>pink-purple pigment</u> at its base and in a few exposed downward-arching roots that give the appearance of a plant "standing on tiptoes." At a glance the <u>ovate leaves</u> appear <u>blunt-toothed</u> but a closer inspection reveals a <u>bristle tip</u> projecting from the underside of each rounded tooth. Veins are <u>pinnate</u>. Lower leaves are <u>opposite</u> but higher leaves become <u>alternate</u>. The giveaway identification clue is to submerge a fresh leaf into water and see it shine as brightly as aluminum foil. <u>Irregular</u>, inch-long flowers dangle from the <u>axils</u> and are uniquely shaped like cornucopia baskets with hooked, tapered ends and <u>petals flaring</u> at the opening. Flowers are <u>orange</u>. Another species (*I. pallida*) produces <u>yellow flowers</u>. Plump seedpods mature with tension in their seams so that they burst apart when touched, springing seeds several feet away. The anti-inflammatory gel in this plant is excellent in resolving topical immune reactions such as poison ivy rash and the itch of insect bites. **Some sensitive-skinned people experience a burning sensation when jewelweed is applied!**

4. **Foamflower** (*Tiarella cordifolia*) has a <u>basal leaf</u> jaggedly <u>heart-shaped</u> due to its shallow <u>palmate lobes</u> and irregular <u>teeth</u>. (It somewhat resembles a red maple leaf.) Its leaf, petiole, and flower stalk show fine <u>hairs</u>. The <u>white flowers</u> have <u>five petals</u>, each very narrow at its base. <u>Stamens</u> extend farther than the petals. From this plant the Cherokees made a medicine to help dissolve kidney stones, a common ailment among their tribe.

foamflower
(Tiarella cordifolia)

5. **Ironwood** (*Carpinus caroliniana*) is a common name given to one small tree in the South (also called "blue beech" or "American hornbeam") and yet to another tree ("hophornbeam") in the North. These trees live together all over the Eastern woodlands and demonstrate how multiple common names sometimes create misunderstandings.

ironwood (Carpinus caroliniana)

The most noticeable trait of Southern ironwood is its <u>smooth gray bark</u> rippled with <u>long striations</u> that resemble muscles and tendons – like the corded leg of a horse. (Note: alder shows this, too.) The trunk is usually twisted. The leaves are <u>alternate</u>, <u>elliptical to ovate</u>, no more than 3" long with <u>pinnate</u> veins and <u>double-serrate</u> margins. On the underside of the leaf, <u>fine hairs</u> follow the veins. The tree was superstitiously believed to impart muscular strength to a Cherokee when he pressed a slab of freshly sliced bark to a self-inflicted wound.

6. **The hemlock tree** (*Tsuga canadensis*) contains no poisons. (**Poisonous hemlocks are herbs, not trees.**) The <u>flat</u> evergreen needles are ½ "-long and grow in what appears to be <u>two ranks</u>, giving the boughs a fan-like appearance. On top of the branchlets grows a <u>smaller version of the needle, approximately 1/3 the size</u> of the dominant ones. These "needlets" are sparsely and linearly dis-

tributed in line with the branchlets and twisted on their petioles so that they lie upside. The bottom sides of both needles show <u>two white stripes</u>, which are rows of pores for gas exchange. Dead branches usually cling to the lower trunk and terminate in very fine branchlets that are among the thinnest found in the Eastern forests. Mostly a tree of the mountains, hemlock can be found in cool coves and ravines of the piedmont. As a shade tree to cool mountain headwaters, it is unequaled. This tree provided the Cherokee with food, cordage, pink dye, fire-kit, kindling, scent-absorber for hunting, and tannin astringents for tanning hides.

7. Jack in the Pulpit (*Arisaema atrorubens*) produces an unusual <u>irregular flower</u> that is sure to catch the eye – a vertical "spadix" covered by a <u>green and/or purple-striped</u>, hooded "spathe." The flower stands erect on a thick stalk, next to one or between two <u>compound leaves</u> of <u>three pinnate leaflets</u>. Each leaflet shows <u>pinnate veins</u>, whose lateral veins connect terminally to form an <u>etched outline of the leaflet margin</u> *within* the actual margin – like a smaller leaflet laid on top of a slightly larger one. When the flower goes to seed in late summer the fruit cluster turns a brilliant <u>red</u>. Also called "Indian turnip," the plant produces an underground swelling (a corm) <u>that must be thoroughly dried before prepared as a food</u>. **If eaten raw, the corm burns the mouth with caustic calcium oxalate**! Indian turnip served as a Cherokee winter staple.

Jack in the pulpit   (Arisaema atrorubens)

Fourth Venture Out into a Hardwood Forest -

1. Violet (*Viola spp.*) – As low plants on the forest floor, violets come in a variety of forms. **Common blue violet** (*V. papilionacea*) has a <u>heart-shaped leaf</u> with small orderly <u>scalloped teeth</u> and a <u>single groove</u> down its petiole, beginning at the upper side of the leaf base. Hairs follow down the petiole. The veins are <u>palmate.</u> (The mid-rib vein and the pair around it might appear parallel at first glance, but see how the other arcing veins do not terminate at the tip.) The network of veins and the slight loft of leaf material between veinlets give the leaf a quilted appearance. The <u>purple flower</u> is <u>irregular</u> with five petals – the side petals being <u>hairy</u> and the <u>hairless</u> lower petal tapering in back to a keel (like the curved bow of a canoe). Flower and leaf are edible. **Yellow violet is somewhat similar but not edible. Experience is necessary before harvesting in the absence of blooms!**

common violet (Viola spp.)

2. **Common Cinquefoil** (*Potentilla simplex*) is a low trailing plant with <u>alternate, palmate, compound leaves</u> of <u>five, toothed, obovate</u> leaflets. <u>Regular yellow flowers</u> from the <u>axils</u> have <u>five petals</u>. The stem is <u>hairy</u>. The leaves are <u>palmate</u>-veined, the leaflets <u>pinnate</u>.

cinquefoil (Potentilla simplex)

pipsissewa (Chimaphila maculata)

3. **Pipsissewa** (*Chimaphila maculata*), also called "**striped wintergreen**," is a low, evergreen herb with <u>whitish, pinnate veins</u> (the mid-rib vein being broad) on its <u>toothed</u>, dark-green, shining leaves that grow usually in <u>whorls</u>. The low stem terminates in a cluster of nodding flowers, each having <u>five white petals</u>. Cherokees used pipsissewa to dissolve kidney stones.

4. **Poison ivy** (*Toxicodendron radicans*) is a vine with many varying appearances and so much misunderstanding attached to it that Chapter 10 is devoted to the plant. Generally, this can be said about its anatomy: <u>pinnate compound</u> leaves are <u>alternate</u> and comprised of <u>three leaflets</u>, whose converging stalks show <u>red-pink pigment</u> at their juncture. The leaflets are usually <u>toothed</u> except where their margins are proximal to neighboring leaflets. Small <u>green flowers</u> grow in <u>panicles</u> from the <u>axils</u>. Vines can be hairy or not. **Contact with any part of the plant, dead or alive, can cause an immune reaction resulting in a rash of itching blisters ... or worse.**

poison ivy (Toxicodendron radicans)

5. **Trillium** (*Trillium spp.*) – There are many trilliums. You'll need a field guide to determine a species. All are erect plants having a single whorl of three large leaves with parallel veins. From the confluence of the leaves emerges a conspicuous three-petaled flower with three sepals beneath. One common species, "toadshade" (*Trillium sessile*), displays mottled leaves and a stalkless

toadshade trillium (Trillium sessile)

maroon flower. Its infant leaves (before unfolding) are edible raw or cooked.

6. **Common Greenbrier** (*Smilax rotundifolia*) is a frequently encountered vine with a four-faceted stem that turns woody with age. It is armed with formidable briers that can be fashioned into fishhooks. The alternate leaves are broadly ovate (often heart-shaped) with conspicuous parallel veins. From the axils arise both curling tendrils (for grasping-support) and umbels of small six-petaled green flowers. Mature berries are blue-black. New succulent spring growth at the tip of a stem – if plump and free of pink pigment – is delicious raw or steamed.

greenbrier vine (Smilax rotundifolia)

wild ginger (Hexastylus arifolia, above; and Asarum arifolia, below.)

7. **Wild Ginger** (*Hexastylus arifolia* and *Asarum canadense*) – There are two entirely different genus classifications that produce American "ginger" plants with usable "ginger" roots (even though they are not related to the Oriental ginger sold in stores). The ginger plant you are most likely to encounter (*Hexastylus*) grows only a few inches off the ground and has a dark green, leathery, heart-shaped, palmate

veined, basal leaf that might be mottled with lighter green. Its ginger smell can be detected from a small tear in the leaf, though the root is the only part to be used. The flowers are shaped like urns and often recline on or under the leaf litter, where slugs and beetles crawl in and out of them and aid in pollination. A common name for the herb is "little brown jugs." The root can be used to dispel nausea, especially from motion sickness. Ginger used regularly in the diet is also known to ease inflammation and lower cholesterol. **Some people show digestive ill effects from eating ginger root! Sample in moderation.**

Blind Medicine Woman – a game to augment plant identification skills

Background: An aged medicine woman – Blue Wolf Woman, the healer of her tribe – is the keeper of plant lore for her village. To ensure that this knowledge is not lost, she has apprenticed a young woman, Circling Otter, to continue her herbal work after she is gone. Just after the girl has learned about plant anatomy (The Botany Booklet, Chapter 2), the chief of the tribe falls ill, and the two medicine women must go in search of the plant that will be used to heal him. During this journey, the old woman is struck by lightning and goes instantly blind and mute.

It falls upon the apprentice to find the plant. Circling Otter stops at every plant and describes some feature of that plant. Blue Wolf either nods or shakes her head. Step by step, the young girl accrues all the facets of the mystery plant's appearance. Finally, at one plant, the girl asks many questions and elicits all nods from her teacher, and she knows that she has found the healing plant.

The Game

Divide your class in half and take each group to a private place with plant diversity. Assign to each group a special plant to learn thoroughly (a different plant for each group) by using their Botany Booklets. Tell each group not to leave telltale tracks, which might give away which plant they gathered around and studied. When both groups feel that they know their plants intimately, tell the students in Group A that they are, collectively, to be the Blind Medicine Woman. They should form a circle with a fifteen-foot diameter (wider if plants are scarce). Somewhere inside their circle is their assigned plant. They stand mute and blind as you guide Group B (collectively, the apprentice, Circling Otter) *carefully* into the circle. Group B must find Group A's plant by asking questions of individuals in the outer circle. *Responses can only be nods or head shakes.*

When anyone in Group A has answered two questions, he sits down and may answer nothing else. This ensures that everyone will be involved in the game. When everyone in the Medicine Woman group (A) is seated, the Apprentice group (B) must come to a consensus and specify the plant by touching it and asking, "Is this it?"

Of course, pointing out the exact plant specimen is not necessary for success, as long as the plant is another of the same kind. If the apprentices triumph in their

plant selection, they remain standing while the Medicine Woman group humbly bows its approval. If the Apprentices are incorrect, assume that they would have harvested a poisonous plant. They must fall down to the forest floor and writhe in agony, as if they had ingested a toxic plant! This bit of comic theatrics helps to assuage the wounded pride that accompanies failure. With this part of the game complete, now the roles reverse, and Group B hosts Group A inside a new circle at the other site.

The questions asked by the Apprentices might go like this: "Is the plant compound? Does the plant have alternate leaves? Are there hairs on the underside of the petiole? Does the stem have three grooves? Does the leaf smell like snap beans?" Etc.

But the questions cannot fall into the realm of what the blind woman could not answer: "Is the plant growing next to this rotten log? Is the plant closer to Annie's side of the circle? Is the plant taller than twelve inches? What does the leaf look like?" Etc.

Tactile Identification of Trees –

When a student feels thoroughly versed in the traits of three specific trees (say, sweetgum, beech, and dogwood), lead the student blindfolded into the forest to examine five trees (three of which are the chosen ones) that you have selected. By touch and smell, can she identify her chosen three? Have students pair off to go through this exercise, then have them reciprocate by switching roles.

At the sweetgum tree she will feel the corky bark that her thumbnail can easily dent, perhaps wings of bark on branches, the star-shaped alternate leaves, and if her fingers are sensitive, the fine rows of teeth along the lobes. If she's lucky there may be a prickly gumball within reach or underfoot. None of these features should be singled out by the guide.

At the beech tree she feels the smooth hard bark, short-stalked leaves with ruler-straight pinnate lateral veins, each of which terminates at a tooth. Perhaps beechnut husks are on the ground. Next year's spear-like buds are present.

With dogwood she can feel opposite leaves, rough scaly bark that rubs off in little chunks. The branch tips may possess an elegant upward curve and an onion-shaped flower bud.

Even if blindfolded students fail, the real value of this game is not in the "testing" portion, but in the intimacy of a solo examination of the Standing People.

The Surest Step to Beginner Plant Identification

Your most accelerated method of learning plants is to be with a teacher who can immediately verify or correct your attempted ID of a plant. Studying alone, you may go for months wondering about a plant's identity as you wait for it to flower. There is nothing wrong with this method, but it's a slow process and you have so many plants to get to know. An experienced plant specialist can settle an identification question for you in a split-second.

Learn about botanical societies, conservation groups, garden clubs, parks, and nature centers in your area to see what they offer in the way of field trips and programs on wild plants. Take advantage of nature trails that offer plant ID tags or self-guide pamphlets. Learn which people in your area are plant aficionados. You'll be surprised how easily friendships can be forged through botany. Go to as many of these programs and field trips as your time allows and ask questions. Ask about the curious plant the leader passed by. Comment on what you have discovered about the plants being talked about. Everyone will appreciate it. Your leader will thrive off your eagerness and in return fill your cup.

<u>*Life-Journal on Plants*</u> – On each plant walk with a specialist, take copious notes. Compile them at home in a notebook, alphabetized by plant name. Keep a running log of the history, identification, edibility, medicines, poisons, and uses of plants. If you find yourself overwhelmed with facts on an outing, concentrate on the portion that most interests you and learn it. Let the rest come at another time. You're still learning at an impressive rate compared to a solo-study method.

But ... your solo time with unknown plants is vital. It is important to feel that you are a part of this process of learning and not merely an audience. Those plant IDs that you earn by your own hard work will remain especially dear to you all your life. Do your solo work first. Then check out your efficiency the next time you are with a knowledgeable teacher.

"In searching for his medicinal plants the shaman goes provided with a number of white and red beads, and approaches the plant from a certain direction, going round it from right to left one or four times, reciting certain prayers the while. He then pulls up the plant by the roots and drops one of the beads into the hole and covers it up with the loose earth."

~ *James Mooney, ethnologist, 1891,* Myths of the Cherokee *and* Sacred Formulas of the Cherokees

CHAPTER 4

How to Interact with Plants

During one of my summer camp sessions, my dozen campers, my assistant, and I were a few miles from our base camp. On our hike we happened upon a garter snake. I caught it for the purpose of giving a snake lesson – pointing out the anatomical traits of a snake that tell you it is non-venomous. (Actually, the garter snake is mildly venomous … but not dangerously so.)

I normally hold a snake with both hands – one behind its head, the other posterior to the anus to prevent it from wrapping around my forearm – but I needed a free hand to point out parts of the snake. I knew what was in store for me, and I was willing to endure it. The snake twined around my arm and smeared its putrid musk from its anus onto my skin – a defensive ploy designed to repel a predator. At this point in the lesson every nose wrinkled as the kids backed way.

When the snake-session was done, I released it and said to my campers, "I need to wash up. Let's get to a shady part of the creek and we'll look for some soap. Then we'll find a sunny place where there are lots of weeds growing so that I can find a deodorant." I wasn't sure anyone had heard my teaser on "soap" and "deodorant." Everyone was frowning at my arm, keeping their distance from me. My arm smelled like it was rotting and would soon fall off.

"Soap?" someone finally asked. Another lesson was just beginning.

At a shady creek bank jungled in rhododendron I spotted the shaggy cinnamon bark and pale elliptical leaves of **sweet pepperbush** (*Clethra alnifolia*). In an atypical moment of quiet, the children watched as I paused, held the plant and thanked it for the leaves that I was to harvest. Dipping water from the creek I scrubbed my

hands together with the leaves. The mound of foam that appeared in my hands was like a magic trick. The kids stared as if I had snatched fire from the air.

"That looks like soap," a boy said.

"It is soap," I reminded.

"But it's just leaves. There's no soap in leaves."

"There is soap in these leaves. Everybody, help me rinse. Cup water in your hands and bring it to me over there away from the creek. We don't want this soap to get in the water."

"How come?"

"It paralyzes fish. If they breathe it, they float right to the top, unconscious. It's one way the Cherokees caught fish – by stunning them with this soap."

sweet pepperbush (Clethra alnifolia)

A dozen skeptical faces stared at me. They brought water in little trickles until I was rinsed. I knew this little bit of suds was not substantial enough to affect the fish in a moving current, but I wanted to squeeze more than soap from these leaves. I wanted a memorable plant lesson. Already a simple bush had these children in awe ... and rightfully so. I watched them eye the pepperbush with furtive deference, as though they were in the presence of a wise elder.

As expected, my cleansing was not complete. The rank smell persisted. This is not a criticism of pepperbush but a tribute to the potency of garter snake musk. In an open space nearby, I found a stand of **white horsemint** (*Pycnanthemum muticum*), which has a particularly strong scent. I said my thanks, crushed leaves, and rubbed them into my arm as the campers awaited more miracles.

"That smells good!" someone said.

white horsemint (Pycnanthemum muticum)

"It should cover up the snake-musk," I said and scrubbed at my offending arm. "It's also a medicine," I added. "Make a tea from this mint and it helps clean out your gut."

"Number two?" someone said. The requisite snickers passed through the group. Then followed the skeptical glares.

"There isn't medicine in leaves," a boy said, frowning. "It's just leaves."

"Actually," I said, smiling, "there is. And it tastes good, too."

"Does it taste the way it smells? Can we make some?"

I raised an eyebrow. "Even though it cleans out your intestines?"

Everyone laughed that off. "If it tastes like it smells …" They were all nodding. I could see their hopes rising. They had been going through beverage-withdrawal in our wilderness outing. No Cokes, no iced tea, no lemonade. Water had been our only liquid refreshment for four days.

"We'll brew up a batch of tea tonight," I promised. "You can drink it around the fire as I tell a story." Twelve smiles greeted that news.

That night we made a kettle-full, and the tea was a big hit. After first rounds, everybody came back for seconds.

"Remember," I reminded, "it's a medicine, too."

The testimonials poured in: "It tastes great!" … "Yeah, give me more!" … "I love this stuff!"

After thirds – even with the story over – they wanted more, but I invoked a law: "No, that's it … no more. Remember, it's a medicine."

Before retiring to our tents, we pored over the topographical map and decided on a destination for the next day's hike – a place called "Cedar Cliffs." From the cliffs we would have a grand view of the mountains to the east. Our destination was only four miles away. We were excited, and everyone went to bed anticipating the trek to Cedar Cliffs.

We never got there. The next morning during breakfast, campers began to drift off toward the latrine. After cleaning up our breakfast cookware, we packed our daypacks and assembled. Everyone appeared ready to strike out.

"I gotta go to the latrine again," someone announced. We waited ten minutes for him, and as soon as he returned someone else said, "I've got to go again, too."

We waited. Upon his return someone else eased out of his pack and let it slide to the ground. There was a look of urgency on his face. "My turn," he said and took off running.

And so it went. We spent the entire morning on the verge of departing but never taking the first step. When it became obvious that we would not have enough daylight to get to Cedar Cliffs and back, we made a different plan for the day – one that kept us nearby the latrine.

That night around the fire, my story began with: "Once there was a group of young explorers who could not believe that there was medicine inside a plant. They made this tasty tea and …"

No tea that night. We made it to Cedar Cliffs the next day. It was a grand view.

The Invaluable Field Guide

Soon we will delve into actually using plants just as the Native Americans did, just as a survivalist might today. We'll consider plants you are most likely to encounter in the eastern United States. The list is by no means complete, but it will provide a wealth of plant usage. If you reside outside this eastern scope, you will find many familiar plants here, but you will need to refer to a book that targets your local range for a more thorough inventory of your area. So, find that book now. No matter where you live, your first job in using plants is to positively identify each and every one you will use.

The Botanical Map – Make a large bird's-eye-view map of your home (or school) habitat, including an outline of buildings, driveways, street curbs, and any other dominant landmarks. Draw a circle of appropriate diameter to represent each tree. Make symbols for large rocks, streams, logs, trails, pooled water, etc. Hang this wall map in a conspicuous place and leave a writing instrument nearby. Throughout the project, as you make positive ID's, write down the exact location of each plant species you discover. Take regular walks through every part of the mapped land, so that you will be sure to catch each plant in its prime – especially when it flowers, for that is the most crucial time for a positive ID. In this way you will know where to find any plant in your neighborhood and learn its guise for that season. Studying the plants that live around you is the best course for a beginner botanist. Learning the plants that grow in other neighborhoods will come in time.

The Scope of Our Plant Work as Practitioners of Survival

In the pursuit of edible wild plants, the first and foremost priority is to identify correctly the species that we intend to harvest. After that all-important step, we must know which part of the plant is usable and at what point in the growing season it should be used. Some plant parts can be eaten raw. Others require preparation. As survival students our work space is forest and field. Our tools and containers are the ones that we make or find; therefore, the procedures we will use are necessarily more primitive than the preparations to which we are accustomed in a home, kitchen, or lab.

Now we should consider foraging ethics. How much of a plant population should we take from a given area? If whole plants are being used, a respectful harvest might be 20% … take one, leave four. Such conservation practices not only protect biodiversity but also set a healthy tone for your mental attitude about the give-and-take of survival.

As we encounter the need for plant medicines, we will tread lightly and create field preparations on the spot. We are not herbalists with a workshop of apothecary jars containing plant materials from around the world. Nor do we have the scales and other measuring devices of the professional. We will use simple techniques with plants that do not pose high risks in preparation.

Plants that contain insecticidal properties (for their own benefit) can often be used as insect repellents for humans, but you must take care to test your skin with a small sample before liberally applying. (This does not mean that you should experiment with any plant as a repellent. You should first learn to identify the insect-repelling plants listed in Chapter 6 to avoid a critical mistake of inadvertently picking a leaf of poison oak, poison ivy, or poison sumac as a trial repellent.) After choosing a plant from the repellent list, crush a piece of leaf and rub it lightly onto a small area of non-callused skin – say, the inside of a forearm. If after thirty minutes there is no adverse reaction, you'll know that the plant is safe for you to use.

All people should not expect identical results with the same plant's insect repellent properties. The effectiveness of the plant depends upon its interaction with your particular body chemistry, which is a product of genetics, diet, hygiene habits, level of exercise and other physical activity, and possibly living and/or working environment.

An overweight person on a strenuous hike might exude scents from skin pores that differ from those of a companion who is in good physical shape. They both might make use of the same plant repellent but the heavier person's emissions might override the plant's toxic deterrent.

Once when I was teaching plants to children in a meadow, gnats began to swarm all over us. The students followed my lead by crushing the disk flowers of **ox-eye daisy** (*Chrysanthemum leucanthemum*) and grinding the material into their faces – especially around closed eyes. As soon as I had finished covering my face with daisy scent, a few gnats flew to my nose and immediately retreated. When this happened a few more times, I knew I had become "gnat-free." The same good results could be seen on a few others in our group … but not for all. Gnats continued to swarm several of the students. Expect these kinds of variations of efficacy.

And finally, we must adopt an attitude of stewardship toward the land that sustains us. Take great care of fragile stream banks. Tread lightly there. When given the choice, do not harvest plants on the sloped shoulders of creeks and rivers. Gather from the level flood plain instead.

Disclaimer

Most books that share information on foods and medicines derived from wild plants contain a disclaimer … and for good reason. We practitioners of survival skills cannot prescribe medicines as certified physicians do. Nor can we know about readers' particular nuances of health or illness or potential allergies. Foods and medicines don't affect all people in the same way.

That said, I have never suffered – nor have any of my students suffered – negative reactions from any of the foods or medicines whose preparations are contained in this book. Students willingly partook of plants by choice after I had shared with them the history, research, and properties of the plants. **In this book we will address foods and medicines from a survival perspective. We'll cover field preparations I have used for likely survival-scenario problems. Other interesting uses of plants are mentioned simply for your edification, but to utilize them you should work under the care of a professional herbalist.**

After making a positive identification of a plant, using it must be *your* choice. The strengths and dosages found in this book should be considered general guidelines that I have worked out for myself, not for the public at large. I have tried to keep medicinal usage on the safe side and can only assure you that these plants have never harmed me. This book gives warnings in bold type about any useful plants that have proved to be dangerous in some way. It warns about look-alike plants that have been mistaken by beginners. This text does not list <u>all</u> poisonous plants.

Some so-called "remedies" of folklore simply don't work. Their false legacies tend to get passed along again and again through authors who simply repeat what they have read in someone else's book. The remedies that I share are ones I have personally used, or – if I have never had need of that medicine – I have known it to be used successfully by acquaintances. Your particular body chemistry, your fragility by age or illness, your complexion, your history of diet, your hereditary constitution, and your allergies (both known and unrevealed) are factors that might cause an unexpected response to a plant.

With these thoughts in mind, enter the world of plants with prudence. When partaking of a known wild food or medicine or repellent for the first time, always test your body's reaction to it by taking just a sip or a nibble or a dab on the skin. Wait a half-hour to see if any unwanted reaction occurs. Then increase the intake or usage slightly for a second experience. Be cautious and safe. It is better to learn now which plants affect you adversely … rather than when you might need them in an emergency.

The Way

Before the hunt, a Native American often took a gift tobacco and a prayer into the forest for the animal he intended to kill. Such reverence was standard practice. I encourage you to create a personal ceremony, to take enough time with a plant to show that you are mindful of the taking, whether the harvest involves a leaf or branch or root or the whole plant. The way that you harvest defines your relationship with the natural surroundings that you ask to support you.

Many non-native people might feel awkward performing such a ritual. This reluctance is merely a symptom of the distance that has been wedged between us and Nature by our present culture. Once you perform your own ritual, you will feel closer to man's original relationship with the forest. Such ceremony of gratitude is not reserved exclusively for Native Americans. The simple act of acknowledging an appreciation toward plants and animals places you, the harvester, inside the circle of life rather than being on the outside reaching in.

Your ritual does not have to be showy or audible. It is yours to develop and deliver as you will. Experiments have shown that plants respond to our moods, our sounds. The practice of ceremony offers to you a new doorway into the forest, one that might not be otherwise available.

We are working our way toward a list of plants that you can use in a multitude of ways. Your chosen field guides will provide plenty of anatomical traits for identification as well as habitat descriptions, so I have not attempted to duplicate that information, which would have doubled the size of this book. But I have added to this list some plants characteristics that I do not see mentioned in books – details that will be helpful to you in recognizing a plant species.

PREPARATIONS

The foods of the wild must be met on their terms. As you try them, think of yourself in a survival situation. One day when you might put yourself into a survival experience – or if life throws one your way – your appreciation of these foods will be assessed by a new gauge of *practicality* and *nutrition*. It is probably fair to say that a great many Americans eat foods today simply to taste them … not out of hunger, but for pleasure. Fast foods with seductive tastes are seldom salubrious – hence the documented demise of American health and physical fitness.

With this in mind, I ask you to step into the garden of wild foods as an open-minded apprentice. Accept the gifts without comparison to less healthful foods. Because the wild survival foods shared in these pages are not offered with elaborate recipes, they cannot compete with your usual culinary experiences. We won't be deep-frying Solomon's Seal root or preparing pizza with acorn crust and wild leek toppings. Instead, we will be dining more plainly with survival in mind. (Don't get me wrong, some of Nature's edibles will surprise you with exciting flavors.) Learn to let nutritional value and a sense of gift be your priorities in assessing wild foods.

Keep in mind that a forager can sometimes find wild foods that "ought" to be out of season. On many a self-imposed survival trip in autumn, I have found plants sprouting new growth on a warm day as if it were spring time. Why? Spring and fall share an important transitional factor – the duration of daylight hours. During unexpected warm-spells in winter, the same phenomenon can occur spurred by temperature alone, as though plants are sensing an early spring.

Tea

To make a tea from **leaves**, if time permits, it is best to dry them for two weeks in a ventilated place away from direct sunlight and moisture. A protective "umbrella" of sticks, leaves, and bark laid across neighboring tree limbs works nicely as a drying shade. Dangle a bundle of tea leaves beneath this and monitor the drying progress. At night wrap the tea leaves to avoid exposure to moisture. How much of a plant's <u>dried leaves</u> should be used for a cup of tea? You already have a sense of that amount. Think of a store-bought tea bag, which contains a "*teaspoon*." Estimate this small mound of crushed dried leaves in the palm of your hand and use for one cup of water. Bring the water to a boil first then pour this hot water over the plant material in a container. (Do not boil the leaves.) Let this steep for a few minutes. Chapter 16 will teach you how to make a wooden container, heat stones, and bring water to a boil. Until then you might practice making a wild tea in your home. Hang leaves inside from a ceiling to dry.

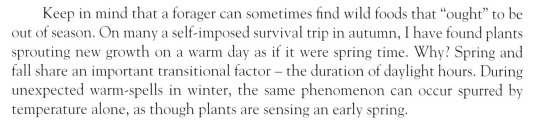

In the field you might find dead leaves naturally dried by sun and wind and time, but these are not useful for tea or medicine. Their exposure to sunlight and/ or moisture has driven out volatile and soluble compounds that are needed for efficacy.

If you have no dried leaves prepared, a tea can be made right away from green leaves. Tear green leaf material into the smallest pieces possible so that the quantity is measurable to the eye. (It would be difficult to equate whole leaves to a teaspoon.) Use twice the volume of green leaves (two teaspoons) and mash the components out of the fresh plant material to coax them into the hot water. (Unlike dried leaves, whose chemicals readily diffuse into hot water, chemicals *inside* green leaves are contentedly saturated in the water inside the leaf and must be forced out with some manipulation.) Constantly crush and stir for twenty minutes.

Some herbal remedies call for a tincture in alcohol. These will not be discussed for survival situations. To further your knowledge in this area, seek out an herbalist.

Roots provide a more potent package of medicinal components. We will discuss only three – sassafras, ginger, and yellowroot – and use special, safe measurements for each.

Survival teas made from the **inner bark** of trees use this ratio: a rectangular outline of the user's little finger defines the amount of flat inner bark immersed in one cup of hot water. Steep for twenty minutes. Always remove bark from a branch (never the trunk) that gets plenty of sunlight. This reduces the possibility of a fungal invasion at the wound. Because inner bark is part of the circulatory system of a tree, we will remove it in the least damaging way.

Always make a linear cut, parallel to the branch's growth, not girdling the branch, which would kill it. The "little-finger recipe" gives you an easy-to-remember quantitative template, but to half its width and double its length spares the tree some critical damage. Some trees can repair severed tubes of inner bark, but most can't.

About Inner Bark and Cambium

One of the most common mistakes I encounter in plant books is the inter-changeable use of the terms "cambium" and "inner bark." They are not the same thing. As described in Chapter 2, cambium is a green, one-cell-thick layer of generating tissue considered to be part of the inner bark. It produces new inner bark conduits. (A second cambium layer can sometimes be present in the inner bark).

When you read that a tree has edible inner bark (maple, cottonwood, beech, birch, basswood, pines, slippery elm, willow) you may be disappointed when you actually try it. Some of the so-called edible inner barks are bitter and certainly not filling. However the cambium layer, especially in spring when the leaves first open, can be more satisfying. At its edible best, cambium exists as a slushy layer that might adhere to the sapwood, or to the inner bark, or to both. Collecting it is a wet-scraping process. So, when you encounter the bitter inner bark from any of the above mentioned trees, if the experience is unpleasant, try scraping up cambium for a sweeter taste.

edible inner bark of pine extraction technique

With pines and their brethren (hemlock, spruce, fir) the entire inner bark can be used but only in moderation due to the harsh terpene content, which gives the bark its turpentine flavor. Terpenes in bulk can irritate the stomach, but they can be cooked out of the bark so that more can be eaten. Still, it's the cambium layer – even with pines – that provides the sweetest taste. [Note: In the above list I used "cottonwood" instead of the proper term "poplar." I did this because too many people believe that the tulip magnolia (incorrectly called "tulip poplar" or "yellow poplar" by so many) is a poplar. It is not. Do not eat the raw inner bark of tulip magnolia. Cottonwood is a common poplar, so I chose it to represent true poplars. (Aspen is another.) Unpleasant bitterness in a wild food should be considered a red flag to your taste buds. Practice moderation to learn your tolerance for such an edible.

Cooking

Over-cooking kills vitamins, but sometimes extended cooking is necessary to make a plant palatable … or safe to eat. Where toxins are not present and cooking is called for, wash the plant and use the clinging water residue to steam the plant. Then test for taste by nibbling. Sometimes repeated cooking (boiling and draining and boiling again) is indeed necessary if the taste is bitter, but the less heat (boiling) we expose the plant to, the higher nutritional value we will enjoy. Do a taste-test when you have poured off the water after each boiling.

In late fall, winter, and early spring harvesting roots might prove challenging to a novice, since the above ground parts of the plant may be desiccated, withered, or absent. Identification becomes more problematic. Harvesting the wrong root could be costly … even fatal.

The earlier *Botanical Map* project will teach you to identify plants in all their seasonal guises … no matter where you may roam. If you learn how to identify the dried weed remnant of evening primrose in winter around the corner from your house, you'll recognize it anywhere in the world.

Chemicals

Plants have evolved mechanisms to protect themselves from animal predation: thorns, hairy textures or tough fibers, and most importantly, offensive chemical compounds – some toxic to humans, some not. Because leaves are a prime food of vegetarian animals, a plant might prudently <u>retain</u> offensive chemicals at the very site of its manufacture (inside the leaf) while its other compounds might flow freely throughout the plant beneath protective outer bark by way of its circulatory system. This explains why a given plant's root, like ginger, might be edible or medicinal while its leaf is not. When a plant's raw leaf can be used by us, it is because there are no adverse chemicals manufactured by the leaf.

Seeds

Seeds – such as from grasses – are an important part of wild foods and should be gathered mature while still on the plant by cutting the seed stalks. Let them thoroughly dry and then shake the seeds from the plant. Rubbing the seeds between the hands helps break up their protective coverings (the chaff) so that the lighter hulls can be winnowed away by wind while pouring seeds from one container to

another on a windy day. Seeds can then be ground into flour using stone on stone, like mortar and pestle … or *metate* and *mano*.

A Brief Introduction to Dyes

The use of plant dyes is not usually considered a survival skill, but color stains can be used to camouflage a primitive hunter's clothing. In addition, using natural colors can symbolize pride in something you have crafted. Dying requires some experimenting and research for each given botanical species. While some plants transfer colors readily to material, others require a mordant or fixer to hold the color to the material. Still, there can be variations of color strengths even among like plants. Some natural mordants include acids from: tannin (acorns, hemlock bark, New Jersey tea root, etc.), grapes and other fruit, sumac leaves, hardwood ashes, red soil, and reddish pond scum. (Store-bought substitutes include vinegar, alum, crème of tartar, salt, and lime.) In general, best results are achieved by thoroughly cleaning the material to be dyed then boiling it with a mordant for an hour. Allow the material to dry. In a large container crush the plant material thoroughly in 2 gallons of water for an hour, then simmer for an hour. Strain out plant fibers, add the material to be dyed and simmer for a half-hour.

<u>Making your first wild leaf tea</u> – Somewhere along an open forest trail or in a field, you are going to encounter one of the mints. Most of these square-stemmed herbs give off fresh, minty aromas when a leaf is rubbed. Find one with a pleasing scent. All the mints have simple, opposite leaves, usually serrate. Flowers are irregular, usually three-lobed on the bottom petal and two above. Flowers grow in the leaf axils or in dense heads.

For a first experience at tea-making, consider the ease of your kitchen, and then later you can approach the process in the wild. Using a knowledgeable friend, then your nose, and a field guide to verify, identify a wild mint and collect mature leaves. There are mountain mints, wood mints, wild thyme, wild basil, horsemint, catnip, bergamots, peppermint and more. **Warning: Two mints – pennyroyal and peppermint – are not recommended for pregnant or nursing women! The juices of horehound, peppermint, and spearmint may cause dermatitis! Concentrated oils of mints are toxic to consume!** Dry the plants for twelve days by hanging the bundled leaves from the top of a doorway in the interior of your house. Boil water. Measure a teaspoon of crushed dried leaves into a one-cup container and then pour the just-boiled water over the leaf material. Steep for two minutes.

To bypass the drying time, use twice as much <u>green</u> leaf material (2 teaspoons of finely minced leaves) and mash it in the just-boiled water with a blunt, smooth implement for twenty minutes.

All the mints work positively on your digestive system, so expect at least some mild laxative action. The cooled tea, applied topically, also has a refrigerant effect on the body.

<u>*Making your first inner bark tea*</u> – Using your new botanical skills, locate a white pine tree (*Pinus strobus*) with a living branch low enough to reach. (See description in Chapter 3.) Lay your little finger lengthwise on top of the limb. Then calculate an outline twice as long and half as wide. A cut along this new perimeter involves the same surface area but inflicts less damage to the limb. Cut this long, narrow rectangle through the bark layers into the wood. (You'll learn to recognize the tougher feel of the sapwood with the point of your knife.)

Don't peel off this ribbon of both outer and inner barks. While both are still attached to the limb, carefully carve away and dispose of the thin, dark-gray outer bark leaving the yellow-white and greenish inner bark. Finally, peel off the inner bark and place it in a mug. Pour in 1 cup of just-boiled water and let steep for twenty minutes. As an option for young students, add lemon and honey to taste. (Pine needles can be crushed into the water also.) This tea is full of nutrition: protein, fat, carbs, phosphorus, iron, vitamins A and C, thiamin, riboflavin, and niacin. Also present is an expectorant to help loosen up congestion from a respiratory inflammation. White pine inner bark was once an ingredient in some cough syrups. **Pine tea should be drunk sparingly to avoid irritation to the digestive tract by terpenes!**

"Each Tree, Shrub and Herb, down even to the Grasses and Mosses, agreed to furnish a cure for some one of the diseases named, and each said, 'I shall appear to help Man when he calls upon me in his need.'"

~ James Mooney, Myths of the Cherokee *and*
Sacred Formulas of the Cherokee

CHAPTER 5
Plants Are Waiting For You

There are people who pay substantial amounts of money to have their lawns sprayed with poisons that specifically target the so-called "weeds" that interrupt the continuity of the grassy carpets surrounding their homes. On the occasion of a bee sting those same people might put a screaming, crying child into the car and rush off to the drug store for a quick-fix medicine to take away the sting. What a shame. It is almost a certainty that broad-leaved plantain had once been available at the doorstep. Relief from the sting could have been only seconds away. And it was free! And perhaps best of all, using that medicinal leaf might have laid down the foundation for a relationship between the child and a plant. As it was, Nature provided only the scenario's adversity – the stinging insect. One of the joys of learning primitive skills is that the outdoors becomes the drug store, the grocery, and the hardware store.

Returning to an Intimacy with Plants

Learning to use wild foods and medicines is not an all-or-nothing proposition. Prior to being thrown into a survival situation, it would be smart to dabble in plant usage. You don't have to abandon your "civilized" way of life and go feral. Complementing your store-bought goods with wild goods is a rewarding lifestyle. Such a realistic approach is enough to keep you in touch with Nature. After all, living completely off your natural surroundings would be a full-time job – just as it would be in a fully-blown survival situation. What percentage of wild food you include in your diet is your choice. Even a sprig of edible weed in a salad elevates an eating experience to an adventure.

Collecting and Carrying

One of the most taken-for-granted implements of modern time is a container. Our culture is saturated with bags, boxes, and other packaging – clutter that we have

been virtually "trained" to throw away. Secretly lurking in most homes are thousands of containers: plastic grocery bags, Tupperware, trash bags, pots, Zip-Locs, pillow cases, pockets, bottles, CD and DVD cases, cans and boxes of foods, trash cans, drawers, purses, etc. One cannot walk an American highway without finding the ubiquitous fast food bags, cartons, and cups strewn across the right-of-way.

In the days of the first Americans, if one found an empty, sun-bleached turtle shell or a sizable seed pod large enough to hold collected items, those natural containers were considered treasures. The value of a tote-device quickly appreciates as a forager accumulates plant specimens. The hands need to be free for work; the supplies need to be contained.

Two trees – basswood and tulip magnolia – were commonly used to make bark baskets for a carrying device. In spring the "green" bark (outer and inner together) can be cut from these trees and shaped into a "berry basket" by carefully scoring, folding, and lacing a single, rectangular slab. The size and shape of the rectangle can be altered to suit different needs: large for a food gathering basket, long and narrow for an arrow quiver, and short and narrow for a dart quiver.

Making a Bark Basket – In the Southeast the tulip magnolia is by far the most abundant source of basket bark. (In the North Country around the Great Lakes, basswood reigns supreme.) Though a living tulip tree cannot replace its excised bark (as, say, sweetgum can), this magnolia seems not to suffer from selective bark removal. In fact, I have seen tulip trees completely girdled by beaver and still survive – a botanical mystery.

In spring cut a perfect rectangle (14" wide X 28" tall) on a smooth, limbless section of trunk. Your knife should penetrate both bark layers and slightly into the sapwood. Due to the tree's vertical grain, all horizontal cuts will require more diligence than vertical cuts. Make especially clean cuts at the corners. Before removing the slab, draw a horizontal midline using a piece of charcoal from your fire. Using this line as a longitudinal center, draw a narrow football shape as shown in the illustration, being sure to taper the ends on a patient curve … not abruptly. (If the ends are bluntly tapered … or if the two curves do not meet precisely at the border of the rectangle, the basket will split or show a "leak.")

Score the outline of the football by cutting into outer bark only. This score-line is where the bark will fold on a curve, giving the basket its three-dimensionality. For a first time basket, it usually takes several attempts at the scoring before the bark will bend without splitting. Remove the slab from the tree by slowly but firmly pushing your fingertips under the bark, palm-side to the slick sapwood. (Be wary of very sharp, aborted twigs that may rise from the wood.) If the bark resists at any border of the rectangle, don't force it. Revisit that border with your knife. When the bark is free use a fist for leverage as you gently bend the bark around your knuckles at the "football." Once both curves bend, the sides of the basket come together, where they can be laced by green hickory inner bark. We'll cover this lacing below.

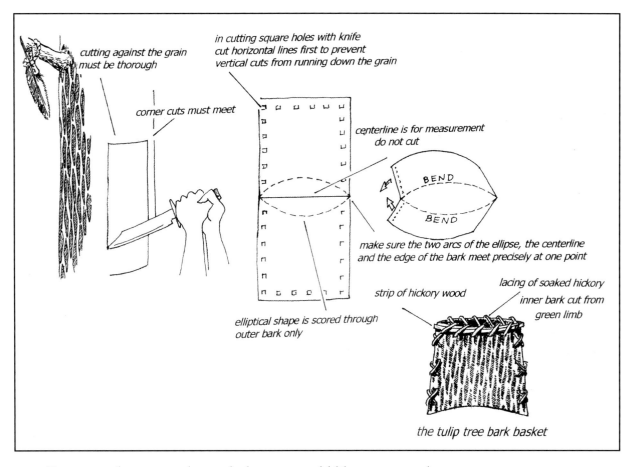

cutting against the grain must be thorough

in cutting square holes with knife cut horizontal lines first to prevent vertical cuts from running down the grain

corner cuts must meet

centerline is for measurement do not cut

BEND

BEND

make sure the two arcs of the ellipse, the centerline and the edge of the bark meet precisely at one point

elliptical shape is scored through outer bark only

strip of hickory wood

lacing of soaked hickory inner bark cut from green limb

the tulip tree bark basket

Determine how many lacing holes you would like to secure the seams, return the slab to the tree (now your workbench), and cut squared holes with a narrow knife blade at least ½" from the edge of the slab. (For each square hole always cut the top and bottom horizontal lines first, then make sure vertical lines do not run outside the horizontal.) Holes must also be cut below the rim of the basket to lash down the rim-stay, a wooden "belt" that keeps the basket's mouth from warping. When this basket naturally cures, it will feel as strong as wood.

Preparing the Basket Lacing and Rim-Stay –

Sacrifice a 4'-long, green hickory branch that shows little or no sub-branching and knots. Carefully carve away and discard the outer bark to cut long, ¼"-wide laces of inner bark. Pull up and soak the laces overnight to get them supple.

Using a vine, measure the mouth of the basket to be made. From the debarked hickory limb fashion a rim-stay (½" wide, 3/16" thick) that will rein-

force the mouth like an exterior belt. A good technique for rendering this stay is to use a mallet (heavy stick) to hammer the back of your sheath-knife through the hickory section as it is held vertically on top of a log (workbench).

Making a Lid for a Container – To create a primitive screw-top lid (for, say, a blow-gun dart quiver) make a narrow bark basket (about 10" X 3") or one that leaves less than an inch of dart protruding from the mouth. Make another mini-basket only 3" tall and slightly wider (by 1/4") than the first. Lace the sides of both baskets but do not add rim-stays. When the bark warps and buckles over the next weeks you will likely find an orientation whereby you can fit the wider basket mouth over the other. (By rotating the cap you will encounter a tight spot that will hold the lid in place, thereby providing full protection to the contents.) Adjustments to tightness can be made by soaking the lid and re-lacing it more loosely or more tightly.

Now it's time to get familiar with a few simple "wild remedies." Here are seven categories of common outdoor discomforts that need a solution: 1.) an itch, 2.) an upset stomach, 3.) a headache, 4.) a cloud of mosquitoes with an interest in your blood, 5.) an insect sting, 6.) intestinal pain from eating food past its prime and 7.) hunger. Each of these problems is listed below with a plant-remedy that I have used with success.

First, read through each section, and then devote a few Saturday mornings to finding all seven plants in your neighborhood so that you'll know where to go should you decide to harvest one. Take your child with you and think of this search as a treasure hunt, for that is exactly what it is. Browsing through a drug-store/grocery store for seven comparable remedies, I tallied up $42 for purchases of small quantities of medicines and food that relate to these topics. (I priced three potatoes, which would approach the nutritional value of one serving of the edible plant listed.)

Of course, buying these products would offer more uses than this one episode of need, but consider how many times we throw away expired medications or food past its prime. One of the advantages of harvesting medicines and foods from the wild on a need-to-use basis is that the material is always fresh. The natural world, it turns out, is a perfect storage facility.

7 Plants for 7 Specific Problems

The plant descriptions that follow apply to their appearances in the growing season and fruit-setting season, spring through early fall. Except with evergreens, identifying plants in winter is a more advanced proposition and is presented separately in Chapter 11 of this book.

You should use discretion concerning the area from which you harvest. Do not use plants growing beside roads or near businesses with suspicious effluents or

where poisons are sprayed. Wash plants thoroughly after harvesting. The plant-uses that follow should be undertaken with a plant identification book so that you can see related or look-alike plants that might be mistaken for the one you seek. A plant ID book is standard freight in the backpack of a student of survival.

1.) **An Itch** – In every plant class when I have led adults and/or children to **jewelweed**, I asked the group if anyone was currently suffering from a chigger or mosquito bite, poison ivy, or any kind of itch. In almost every case, someone raised a hand. After an introduction to the plant, the afflicted one applied jewelweed's juices, and in every case the itching stopped *immediately*. (See description in Chapter 3 and photo in color section.)

Stopping an Itch

Stopping an Itch – In summer look along the sunny edge of a lake, stream, marsh, pond, water ditch, or other low wet ground, where plant growth is lush in healthy soil. Search for jewelweed (*Impatiens capensis*), also called "touch-me-not" for its "exploding" seed pods.

For poison ivy rash, mosquito or chigger bites, or fire ant stings (and other topical complaints), crush part of a stem and its leaves and rub the mucilaginous juice into the afflicted area. This gel contains an anti-inflammatory compound that works *immediately* to relieve itching. The stem juice is readily available up to late summer when the plant hardens. At that point, crush multiple leaves to apply gel. **People with sensitive skin might experience topical burning from this plant juice!**

2.) **Upset stomach** – Yellowroot (see Chapter 3) contains berberine, a compound soothing to the mucosa. (It resolves nausea but is not effective on stomach virus.)

Relieving an Upset Stomach

Relieving an Upset Stomach – I dig for one small piece of root from brilliantly colored **yellowroot** on a flood plain (sparing the more fragile creek bank) and cut off a piece equivalent in size to ¼" of packaging string. After gently washing away any dirt I chew the root for several minutes, swallow the bitter juice, and then spit out the woody fibers.

3.) **A migraine** – On nine different occasions students have come to me during a class to inform me about the onset of a migraine headache. In obvious discomfort each said she needed to go home to lie in the dark in complete quiet for several days.

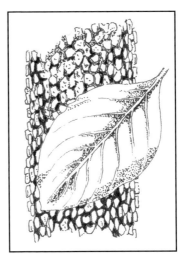

Each time this happened I asked if she would like to try a tea of inner bark from flowering dogwood (*Cornus florida*). The result was successful in every instance.

On the upper side of a branch I cut a long thin rectangle based upon *her* little finger, carefully sliced away the outer-bark, and peeled up the exposed rib-

bon of inner bark from the sapwood. I put this strip of inner bark into a container and over it I poured boiling water. After the bark steeped for twenty minutes, she drank the tea and lay down for 45 minutes, after which she was symptom-free and once again participating in the class. As far as the other students were concerned, I couldn't have asked for a better lesson about dogwood. This same flowering dogwood tea also reduces fever. **Herbalists warn that dogwood inner bark must be dried before using, so as not to upset the stomach! The dosage of "green" bark suggested here has not, in my experience, created any adverse effects; but it may be wise to start with a smaller dosage in case you are sensitive to cornic acid.**

Easing a Headache

– Because non-migraine headaches are more common, we will use a willow tree to resolve "chronic" headache. Black willow (*Salix nigra*) makes salicylic acid (related to aspirin's acetyl-salicylic acid) and sends this chemical into its inner bark. I simply remove the leaves from the outermost inch of a green branch tip and then chew the succulent twig for a few minutes. I swallow the bitter liquid and finally spit out the fibrous pulp. It takes 6-8 hours to be metabolized, making it more suitable for long-term problems. (Photo in Chapter 3.)

4.) **Mosquitoes** – On one of the Golden Isles of Georgia, a group hired me to run a symposium on tracking. It was a perfect request, for tracking on sandy ground opens up endless teaching and learning opportunities, especially on an island rich in wildlife.

On one of our hikes, a cloud of big, hungry, persistent mosquitoes hovered around us and sank their mouthparts into every bit of exposed flesh. There were children and adults present, and soon everyone was either whining, in tears, or bitterly angry at the unrelenting siege. Some of the men were slapping themselves so hard that their skin reddened almost to the same hue as the blood that beaded on their bites.

I carried a small vial of homemade repellent in my first-aid kit and assumed that everyone in the group would have brought a repellent, too. Out of twelve people, only one had packed a spray repellent, but after pumping it a few times, the bottle was empty. My supply was not enough to go around, so I was on the lookout for a plant that might afford them some protection.

bracken fern (Pteridium aquilinum)

We were quite a few miles from our base camp, when I spied a colony of **bracken fern** (*Pteridium aquilinum*) in a pine forest. In the mountains I am accustomed to finding bracken in dry, sunny, "waste" places of poor, acid soil, but I have encountered this fern in moist forests, too. Here we were in alkaline soil. You never know where you are going to encounter bracken.

We rationed our treasure – two plants to a person – leaving plenty growing under the pines where we had found them. After testing everyone's tolerance to the plant, we gently rubbed the ferns between our hands to release their insecticidal chemicals and then lightly brushed them on our arms, necks and hands. Finally, we inserted the ferns under our hats – one in front and one in back, like the double bills of a deerstalker cap. Those without hats made a length of cordage from palm fibers to serve as a headband.

As we continued our trek toward camp, we still itched from our previous bites, but all the talk was about the invisible shields that we wore. There were no more bites.

Repelling Mosquitoes

Repelling Mosquitoes – Bracken is a large coarse fern that grows in a variety of habitats that differ in shade, moisture, and soil types. Its leaves subdivide into *leaflets* and *sub-leaflets*. The leaflet <u>tips</u> lose their laciness and appear melded together. All the leaves together make a rather horizontal triangle that might stand as tall as your knee or shoulder. [Compare bracken to its lacier look-alike rattlesnake fern (*Botrychium virginianum*.)] Pick the entire plant and lightly bruise the greenery between your palms. Place the fern partially under a hat so that the foliage projects over your face like a hat bill.

5.) *Insect Sting* – During a grammar school presentation on medicinal plants, I took a fourth grade class outside to see what treasures were hiding on the school grounds. The first plant we talked about was **broad-leaved plantain**. We discussed in great detail how to recognize it and how to use it as an effective poultice on insect stings. (See description/photo in Chapter 3.)

We were nearing the end of our time together when the teacher of the class emitted a sharp chirp. Everyone turned to see her slapping repeatedly at a wrist. She had been stung by a fire ant.

"We'll need to go back inside now," she said angrily. "I need to get something on this."

"What will you put on it?" I asked.

A little impatient, she shook her head. "I don't know. Something. Some antibiotic ointment maybe."

"It stings because the ant injected formic acid into you," I reminded. "It's an acid burn."

"Whatever," she said and began herding the children toward the school building.

"How about we neutralize the acid?" I called out, hoping this might jog someone's memory. Everyone stopped. She gave me a semi-tolerant smile.

"Well, how in the world would I do that?"

I looked at the children and waited. Finally a light bulb flashed over several young heads.

"That leaf with the strings inside!" a little boy said.

Smiling, I pointed at him. "Plantain," I reminded.

We returned to the plantain, crushed a leaf, and the teacher rubbed it into her hand. The kids watched, riveted to her grimace. In less than a minute she locked eyes with me.

"It's stopped hurting," she said a little surprised.

Our lesson had transcended from an academic session of weeds and words to a personal experience with a problem and its solution. At least in the teacher's mind, our original lesson with the plantain had not been relevant to her. It never occurred to her to use the very plant that had begun our lesson as a remedy for insect sting.

We had begun strolling back toward the school building, when she stopped and looked back to the area where we had found the plantain. "Let me go look at that plant again," she said. As we walked back to it, she asked, "Now, what was it called?"

Neutralizing Insect Venom – Once called "white man's foot," broad-leaved plantain (*Plantago major*) is not native but is now widespread throughout America. It is common on hard-packed dirt (trails) and in yards where direct sunlight is ample.

If a bee or wasp's stinger is lodged in the skin, scrape it away without deflating the venom sacs, which are still attached. Then crush a plantain leaf until it becomes juicy. If you are in a pristine area, chew the leaf into a pulpy mass. Press the crushed leaf directly on the sting-site and rub. After a minute throw away the leaf, harvest another, and then repeat the process.

6.) *Tainted Water or Food* – On a self-imposed survival trip I had the misfortune of ingesting microorganisms that wreaked havoc on my gut. So as to spread out my bounty over time, I had eaten portions of the same cooked fish over a 24-hour period. Despite my efforts to keep the meat cool in a stone "refrigerator" built over the creek, by the second day the fish tottered on that border between edible and starting to spoil. I should have discarded it, but having taken the animal's life I was reluctant to throw it away. Within minutes after another meal of fish, I was so ill that I could not perform the simplest of chores.

As a precaution, before I'd eaten that "last supper," I'd located a **sassafras** tree, dug up and cut off a piece of root, and brewed a tea. Later, in my misery, I drank one cup and lay down. I'd recently learned about the research which revealed that the constituents of sassafras root killed harmful microorganisms in the human gut. My experience was a most impressive substantiation. Less than an hour after finishing that cup of tea, I was symptom-free and back at survival work.

To make this tea I dug with a digging stick around the tree's trunk, located a root, and followed it out to a fork to remove the smaller branching rootlet. For one cup of tea, I used a piece of cylindrical root ¼" thick and ½" long. Back at my camp while waiting for water to boil, I washed the root and cut slits into it. Then I poured a cup of just-boiled water over the scored root and let it steep until the water turned a faint pink – about fifteen minutes. I removed the root and wrapped it up for future use. **Refer to Chapter 6 for the Surgeon General's warning about consuming sassafras.** Note: Herbalists usually <u>boil</u> roots for an extended period. This steeping of the root worked well for me. Sassafras tea should not be used regularly or as a prophylactic measure for questionable water. It destroys not only bad bacteria in the gut, but also the good.

<u>Getting to Know the Sassafras Tree</u> –

Locate a sassafras tree (*Sassafras albidum*), often in a dry hardwood forest, by spotting its three (or more) differing leaf shapes: an elliptical football, a mitten, and "Casper the Friendly Ghost." Be certain that the leaf margins are smooth so that you do not mistake mulberry (*Morus rubra*) for sassafras. Scratch away a flake of outer bark and beneath it you'll see a cinnamon color. To verify identification, bruise a leaf and smell the flavor of *Froot Loops* breakfast cereal or *Davy Crockett bubble gum* from the '50's. (Sometimes poor soil conditions make this scent very faint.) Dig close to the trunk to locate a lateral

sassafras

sassafras tree (Sassafras albidum)

root and check its identity by smell. A freshly cut root will present the unmistakable aroma of root beer.

7.) *Hunger* – On my self-imposed survival trips in fall and winter, I have always felt especially indebted to **groundnut** (*Apios americana*) for its sheer volume of food. I never miss an opportunity to explore a sunny thicket near a creek or the moist edge of a low meadow.

Not only does groundnut grow in a colonial fashion (if you find one, you're going to find more nearby,) but also a single root might produce five or six tubers – like beads strung along a bracelet. Baked in the coals of a fire, these leguminous tubers pack 3-times the protein of potatoes.

Eating a Wild Tuber – Get to know groundnut in the late summer, when

its leaves and flowers make identification definite. The leaves are pinnately compound with five to seven ovate leaflets. Often the leaf appears semi-folded along the rachis, like a closing book. The brown-purple flowers, which cluster in racemes, show the classic irregular shape of other pea flowers. Mark the tiny vine with a ribbon so that you can find it when the leaves have dropped. Do this enough times until you can recognize the leafless vines in winter. Fall or winter harvest ensures that tubers are packed with nutrition (from the summer storage).

The groundnut vine twines in a delicate spiral, often around desiccated stems of goldenrod or other tall weed remnants. <u>Very carefully</u> follow the fragile stem to earth and dig, following the root without breaking it, to find several tubers spread out on the same root – sometimes close together, sometimes not. The tubers are often golden tan and warty, varying in size from bird's egg to softball or larger. You'll see bitter latex oozing out of any breaks in the peeling. To dispel bitterness cook the tuber in any of the many ways you might cook a potato.

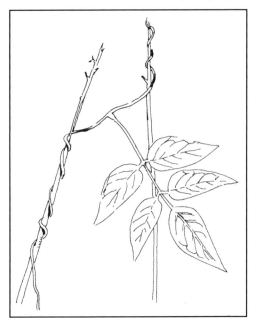

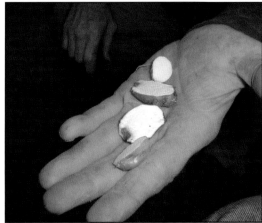

groundnut (Apios americana)

"If the plants of the forest are the work of the Great Spirit, are they not holy?"

~ *Crow Littlejohn*, Requiem to the Silent Stars

CHAPTER 6
100 Plants
~ And Their Many Gifts ~

<u>**X 100 Plant Adventures**</u> – Consider each of the more-than-one-hundred plants listed below as a dual adventure – a treasure hunt followed by a project to make use of the found plant … if there is a need. First browse through the list to choose a plant that interests you … or one that you suspect grows nearby. For safe harvesting use your chosen botanical field guide to find and positively identify the plant. (Plant photos are found in the color section at the end of this chapter or in Chapters 2-5, where a plant has been previously discussed.) This sampling covers edible, medicinal, and craft uses of plants commonly encountered in southern Appalachia (by no means is this a complete list), but many of these plants can be found all across the country. Because you now should have in hand a plant identifying book, I have not included exhaustive physical features for each plant, but I have entered anatomical "secrets" not mentioned in other books.

Using your self-made Botany Booklet, make notes on your observations of a plant's anatomy and compare your own assessment of the plant to descriptions in a field guide text. If a plant's "survival uses" intrigue you, take on a project. **Some medicinal plant uses that fall outside the scope of a survival scenario are mentioned here as points of interest and marked:** *"Other applications to explore with an herbalist:"*

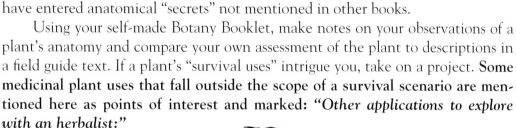

<u>AGRIMONY</u> (*Agrimonia sp.*) – This herb catches the eye with its orderly alternation of small and large leaflets. A tea of summer leaves applied topically to wounds is antibiotic, anti-inflammatory, and astringent. It can be used as a gargle for sore throat.

<u>AIR POTATO</u> (*Dioscorea batatas*, also called "cinnamon vine") – The air tubers of late summer – which look like miniature Idaho potatoes – are edible raw or cooked. The underground root can also be cooked and eaten. **Do not confuse with morning glories and their relatives!**

ALUMROOT (*Heuchera americana*) – In appearance this herb strongly resembles foamflower but prefers rock outcrops for its habitat, while foamflower prefers moist stream banks. Flower arrangements make the distinction easy (see photos in Chapter 3 and color section). A leaf tea can be sipped for diarrhea. Root tea (one plant's root system in one cup of water) can be applied <u>topically</u> as a styptic for minor cuts. The powdered, dried root can be sprinkled on a larger wound to stanch bleeding.

AMARANTH (*Amaranthus spp.*) – Commonly found in abandoned gardens, young greens can be eaten raw or cooked as a potherb. The small dark seeds can be dehusked by rubbing, separated from chaff by winnowing, and eaten raw or roasted or ground into flour. Native Americans farmed this plant as a staple for bulk supplies of high protein bread flour. **Plants growing around barnyards have been noted to accumulate harmful concentrations of nitrates!**

ASH (*Fraxinus spp.*) – One of the few opposite-leaved trees, this one pinnately compound. Its young bark is often warty. Older bark ridges make diamond-shapes. In summer a tea made from crushed green leaves can be daubed on the skin as an insect repellent. Apply sparingly to check for your skin's particular sensitivity to it. Dead ash wood makes a good bow-and-drill fire-kit.

AUTUMN OLIVE (*Elaeagnus umbellata*) – This introduced tree/shrub has silvery, reflective undersides of leaves. Because forestry managers made use of this alien tree in wildlife openings, it is now quite common and considered invasive. Birds and other animals have spread the seeds and expanded the tree's range, crowding out native plants. In high summer the edible fruit ripens to dull red with beige or gray spots. It is our richest source of the antioxidant lycopene.

BAMBOO (*Bambusa spp.*) – Known as "the plant of a thousand uses," most species of this introduced grass are hollow except at the septa (the walls between sections). Canes are used for construction (scaffolding, shelter frame, furniture), cup, flute, fishing pole, bow, blowgun, pipe-work as a water conduit, fences, fish trap, basketry, fire-making, digging stick, knife, spear, arrow, and more. Shucked of their papery sheaths, young shoots of some species are edible; **but all bamboos probably contain some bitter precursors of cyanide. Boiling away bitterness is the safest method for preparing as food. Determining bamboo species may not be feasible, so let bitterness and discretion be your guide. Rivercane, our native lookalike, is toxic if eaten raw! Green sections of mature bamboo explode when laid on a fire! A half-inch-thick cane sounds like a .22 pistol, two-inch-thick cane like a shotgun. Beware of flying coals!**

BASSWOOD (*Tilia americana*) –This tree was the centerpiece of Ojibwa village construction. (See description in Chapter 3.) Its soft wood fibers provide a prime carving wood. Spring bark can be cut off in sheets for shelter shingling or a basket. The inner bark can be used green "as is" for cordage or stripped from dead wood and woven into rope. In spring the living cambium (a slushy green layer next to the inner bark) is edible as a survival food. The burgundy winter buds are edible and contain an okra-like mucilage that is filling. The tips of young, tender branch shoots and unfurling leaves can also be eaten in spring. A pleasant tea can be made from its dried flowers. Dead inner bark, when frayed and fluffed, makes

excellent fire tinder. The wood is exemplary for fire-kits, either bow or hand drill. In the Southeast, basswoods are usually found individually on a stream bank. Farther north to the Great Lakes they can dominate floodplains, often growing in tight clusters – four or five trunks from the same root system of a deceased tree.

BEAUTYBERRY, AMERICAN (*Callicarpa americana*) – The shrub's shiny pale magenta berries are edible. Crush the fresh leaves and apply to skin as an insect repellent. The oil inside leaves has been found to be as potent as the commercial repellent *Deet*, but re-application is necessary every 2-3 hours.

BEECH (*Fagus grandifolia*) – Beech holds its coppery dead leaves through the winter into spring, making it easy to locate in those seasons. In fall, if the tree produces fruit, two edible nuts (in shells) are found inside a prickly husk – each shell being triangular-shaped like a single-pole tent. Leaves and bark are antiseptic and astringent. Very young leaves are edible. Inner bark is said to be edible, but like other inner bark food candidates, you will find the spring cambium slush most palatable due to lesser concentrations of saponin and tannin. **Note: Beech nuts contain traces of a toxin called fagin and should be eaten in moderation. Fagin is concentrated in the papery husk, easily removed after roasting by an open fire.**

Beech Nut

BIRCH, BLACK (*Betula lenta*) – The roots tend to snake on the surface of the earth for some distance from the trunk, suggesting they cannot tolerate saturated soil. The lenticels (gas-exchange pores on the bark) are pronounced horizontal "dashes." Splotches of light- and dark-gray bark adorn the trunk. Black birch produces a great many alternating dwarf twigs called "spurs." Spring leaves emerge from these spurs tightly bunched in pairs that appear opposite in arrangement, but all birches grow alternate leaves. Scratching the branch (in summer) or root bark (anytime) releases a wintergreen aroma. Though toxic in quantity, methyl salicylate (wintergreen oil) is analgesic and mildly sedative when used in moderation. Because the oil is flammable, dead strips of outer bark make fair kindling, offering a protracted chemical-burn. (The more papery bark of other birches burns much better.) Dead birch wood can be used to make fire by friction; but because persistent sleeves of bark are resistant to rot, the interior wood is often damp and will need drying. In the growing season wintergreen oil can be derived from fresh twigs scored and steeped to make a delicious tea, but drink only occasionally. (Cut slits into a wooden-kitchen-match-sized twig for one cup.) Used topically, this same tea can be rubbed into aching muscles. The inner bark is a survival food to be eaten sparingly – raw, cooked as noodles, or dried and ground into flour. A strong twig tea used topically can ease the itch of poison ivy rash. (See photo in Chapter 2.)

River Birch (Betula nigra)

BLACKBERRY (*Rubus spp.*) – The compound leaves can be palmate or pinnate. In spring tender new shoots are edible before they become bitter. Black fruit is delicious in summer. (Prepare for briers and an abundance of chiggers.) The mature leaf tea's astringency resolves diarrhea. Passable cordage fibers can be accessed by gently pounding mature canes.

BLOODROOT (*Sanguinaria canadensis*) – The pure white of the early spring flower, the ochre-orange sap inside the leaf, and the blood-red root-sap give this plant an array of striking colors. The root-sap can be used sparingly on the skin as an insect repellent, **but some people react to applications with eruptions of dermatitis! Test a small area of skin before applying. Bloodroot is a protected plant and should be utilized only in emergency.**

Other applications to explore through an herbalist: The concentrated root sap corrodes human tissue and has been applied to skin fungus (ringworm), surface cancer, and polyps. The Cherokees, who were susceptible to a fast-spreading skin cancer, applied the root juice topically, where it destroyed cancer cells and repaired DNA in others.

BLUEBERRY (*Vaccinium spp.*) – Lowbush blueberry is common on dry mountain elevations where it grows in colonies. Highbush blueberry needs moderate to good sun and is often found on lower ground. Look for the small "crown" on the underside of dark blue, ripe, edible berries. The blue-green fruit of deerberry (*Vaccinium stamineum*) is most often too sour to enjoy.

BRACKEN FERN (*Pteridium aquilinum*) – This large lacy triangular-shaped foliage [learn to distinguish it from Rattlesnake Fern (*Botrychium virginianum*)] forms colonies in summer in dry or wet areas. For an instant insect repellent, bruise a frond, rub into clothing, and finally wear it like a visor to protect the face from mosquitoes and other insects. (See photo in Chapter 5.)

When bunches of harvested bracken are stacked together they make a comfortable loft for bedding and provide an insecticidal barrier from the ground. **All parts of the mature plant are filled with toxic chemicals! See "Ferns" in this section.**

BUCKEYE (*Aesculus spp.*) – This is one of our few native opposite-leaved trees (the only one with palmate compound leaves) with longish, oblanceolate, serrate, and pinnately-veined leaflets. The softness of its dead wood makes it an excellent fire-maker, especially as a hearth for the hand drill. Leaves can be crushed for soap. **Historical note:** Buckeye was a Cherokee sacred wood, carved for ceremonial masks. Though not substantiated by science, buckeye sap was said to be a Cherokee sun-block and a

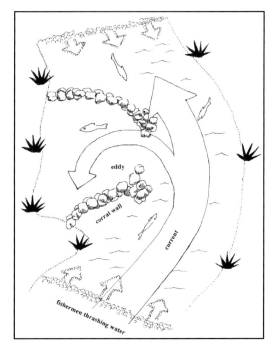

treatment for sunburn. Cherokee women knew how to leach the poison out of the nuts – a complicated process that I purposely omit. (If improperly prepared and eaten, the nut's glycosides destroy red blood cells and afflict nerve tissue.) Crushed nuts are effective as a fish poison. Fish were herded into a rock corral by flanks of people walking in a stream. The corral was closed off, and nuts were thrown into the calm water area. Stupefied fish floated to the surface and could be picked up by hand. Fish-poisoning is now illegal.

<u>BUGLEWEED</u> (*Lycopus uniflorus*) – A creek, pond, and swamp dweller, this mint is not noticeably aromatic. Learn to identify the winter remnant of its desiccated stem. Its underground tuber lies close to the surface and is edible raw or cooked from fall to early spring.

<u>BULRUSH</u> (*Scirpus spp.*) – (See cattails for identical usage of all parts except flowers.)

<u>BURDOCK</u> (*Arctium spp.*) – The deep, first-year plant's taproot can be cooked until tender enough to eat. (Experiment with different sections of the root.) Young leaf stalks can be peeled and eaten raw. The young leaves need boiling once or twice but might never be palatable. To eat the young flower stalk (still supple before flowering) scrape away its rind to a white core and enjoy raw or cooked. (Photo in Chapter 5.)

<u>CATTAIL</u> (*Typha spp.*) – This plant is an important find, because it offers food during every season. In spring the new shoots emerging from water can be peeled of outer leaves to get to the delicious soft white core ("heart of cattail"), which can be eaten raw or cooked. On a tall stalk both flowers (male above, female below) develop inside a sheath. (It is the female flowers that will eventually resemble a "hotdog on a stick.") The male flower can be steamed and eaten like corn on the cob. The pollen produced by the male flower is copious (invert the flower and shake inside a bag) and can be eaten raw (mildly sweet) or used as flour for flat bread or soup thickening. Running through the mud connecting individual plants, the rhizomes (underground stems, most with a tan leathery covering and some smooth white) are full of starch that, in the cold months, can be eaten raw from the interior. **Take care that your water habitat is not polluted and that you have thoroughly cleaned or cut away all mud. (Reports of sickness may be indications of neglect in these areas or from eating rotten parts. It is safest to bake, boil, or roast the rhizomes.) Also present in the rhizomes are many strands of indigestible fibers. Spit them out!** Cut open a cooked rhizome and chew the interior mass for its soft starch, swallow the liquefied starch but not the tough matted fibers. (Those fibers provide a source for rayon.) At the ends of some rhizomes or at the swollen base of the plant look for a new shoot in the shape of an animal's horn. This "horn" is edible after peeling away its layers of protective sheath. The gooey gel between mature leaves has been documented as a topically-applied blood coagulant and local anesthetic, though I personally have not had impressive results with it. The dry, gone-to-seed female flower – when twisted in the hands – provides a surprising amount of seed-down for insulation material that can be stuffed into clothing in cold weather … or used as comfortable padding, a diaper insert, or fire tinder. Soak long leaves and weave into mats or bedding. Desiccated brown leaves can be

rehydrated and made into cordage. Robust stalks can be used as a hand drill for fire creation.

CHERRY, BLACK (*Prunus serotina*) – At the southern end of the Appalachians these edible, tart purple-black cherries ripen in late spring … later in the north. Recognize the tree by its extremely rough and curled, gray to black, thick, bark plates and the presence of a jelly-like substance that hardens to black crusty globs ("black knot fungus"). Dark green, serrate, leathery, elliptical leaves often host small standing pink "towers" of insect galls on the upper surface. **Wilted leaves produce cyanide! Smell this poison by crushing a mature leaf and waiting several minutes. Cyanide poisoning can be fatal! Historical note:** Cherokee women knew how to render the toxic seed pulp edible, but the work is extremely tedious and time-consuming. Like apples and peaches, cherry pits contain cyanide-related toxins. ***Other applications to explore through an herbalist:*** Inner bark tea can be prepared as a sedative for coughs.

CHUFA (*Cyperus esculentus*) – Look for this sedge in moist meadows, ditches, and edges of ponds and streams. Atop the 3-sided stem grass-like leaves spread from a whorl below a cluster of feathery flowers. Underground tubers, sometimes found in loose sandy soil, are edible raw, boiled, or dried and ground to flour. For a coffee substitute, roast the tubers until brown inside and out and use like coffee beans. A lookalike is tall umbrella sedge, which grows no tubers.

CLEAVERS (a bedstraw also called "goose grass") (*Galium aparine*) – Whorls of leaves grow on a soft square stem. The barbs projecting from stem and leaves make this succulent plant so "sticky" that a piece of the herb thrown onto cotton fabric will "stick" in place. The seeds can be roasted for coffee. Young tender shoots can be thoroughly cooked as greens. Leaf tea is applied topically as a cool wash on sunburned skin. **Historical note:** Because of its spongy construction, cleavers was once used by pioneer women as a goose down substitute to stuff mattresses. (See bedstraw photo in Chapter 3.)

CLOVER (*Trifolium spp.*) – Traditionally, flower heads and young leaves have been eaten after soaking in salt water for 3 hours or boiling in fresh water for ten minutes. These same parts of the plant have been dried and ground into flour. But **there are cases of clover being toxic or infected with a disease, neither of which can be detected by the eye! For that reason I cannot recommend using!** (Over-eating even healthy clover can cause bloating!) ***Applications to explore through an herbalist using safely grown clover:*** Clover contains estrogens and genistein, a chemical that appears to trigger the immune system to stop the spread of breast cancer. The flower tea shows sedative properties. (Photo in Chapter 3.)

COHOSH, BLACK (*Cimicefuga racemosa*) – The stem tri-forks into twice-compound leaves. The white flowers can be rubbed between the hands and applied to the skin as an insect repellent. **Historical note:** The knobby root gives this plant its native name, which means "rough." ***Other applications to explore through an herbalist:*** **Blue cohosh** root tincture is commonly used by midwives to initiate childbirth, while **black cohosh** (a central nervous system sedative that has proved to alleviate the pain of childbirth) helps a new mother's body return to normalcy.

CORN SALAD (*Valerianella olitoria*) – Common in unattended gardens, corn salad offers edible, tender, young leaves. Harvest before flowers form and eat raw or cooked. It is high in boron (which aids in calcium absorption) and in tryptophan (the precursor of serotonin).

DAISY, OX-EYE (*Chrysanthemum leucanthemum*) – The young stem, leaves, and flower buds can be nibbled or added to other foods as a tangy flavoring. Though thin and tasteless, mature ray flowers are edible, too. Mature yellow disk flowers (the "sun" of the "day's eye") can be crushed into clothing and skin to repel insects.

DANDELION (*Taraxacum officinale*) – Though bitter, young greens are highly nutritious eaten raw. Tastier blanched leaves are often found beneath sunnier leaves of the basal rosette. Boil, bake, or sauté flowers, even as buds. Collect the root in the fall, dry, and roast for a coffee substitute. ***Other applications to explore through an herbalist:*** Fresh root tea makes a smart diuretic, because it is high in potassium – one of the electrolytes a human needs to replace when losing fluids. Boil flowers for a yellow dye; roots for a magenta dye. (Photo in Chapter 3.)

DOCK, CURLY (*Rumex crispus*) – Very young dock leaves (folded or still tender and pliable), petioles (plump and purple pigment-free), and young flower stalks can be eaten raw (or boil for better taste). Due to high levels of oxalic acid, consume in moderation. If your body tends to make stones, abstain altogether. Older leaves might be cooked in two changes of water, but the loss of nutrients (calcium, potassium, phosphorous, protein, iron, and vitamins A and C) probably negates the effort. The juice of a crushed mature leaf topically neutralizes nettle and fire ant stings and is anti-fungal to use on ringworm. Simply rub it into the afflicted area. The dried crushed root is antiseptic and styptic for bleeding wounds and as a paste for sunburn. See Chapter 3 for descriptive details and photo.

DOGBANE (*Apocynum androsaemifolium*) – Toxic stem fibers make a polypropylene-like rope.

DOGWOOD, FLOWERING (*Cornus florida*) – Called "alligator tree" by the Cherokee, this dogwood usually demonstrates bark broken up into blocky patterns like a reptile's scales. It is one of our opposite-leaved trees. Inside the mid-rib vein is a fragile latex thread that can be exposed by gently pulling the leaf apart. (A similar thread can be found in black gum, silverbell, persimmon, and a few other tree leaves.) Note the pinnate veins that curve away from the margin (called "arcuate" veins). The inner bark contains cornic acid, an aspirin-like painkiller and quinine substitute. ***Applications to explore through an herbalist:*** A tea of the inner bark reduces fever. In my experience with students, it has resolved a migraine within an hour. (Refer to Chapter 5.) **Historical note:** Dogwood inner bark was used by the U.S. military in the Civil War and during past malaria outbreaks for its fever-reducing capacity when quinine supplies were depleted.

DUCK POTATO (*Sagittaria latifolia*) – The large arrowhead-shaped leaf has a palmate venation with distinctive parallel veins running out into each of its three lobes. Called "wapato" by the Native Americans, aquatic duck potato produces delicious underground tubers in its quest to asexually reproduce. From early fall to spring locate a rhizome in the mud of a stream or pond bed and follow it to a

terminal swelling. When popped free, this tuber floats to the surface. Steam, boil, or bake. Young new leaves (still furled) and their stalks can be boiled/steamed and eaten. The young emerging flower stalk, while still quite tender, can be prepared in this same way. **Do not confuse with arrow arum, also palmate but with pinnate veins in the lobes!**

<u>ELDERBERRY</u> (*Sambucus canadensis*) – This shrub has opposite, pinnate, compound leaves. Flower petals (white) are edible. Purple-black berries in summer are edible and high in vitamins C and A, potassium, and iron. **(Several varieties of elderberry exist, some that ripen to red. Reports of upset stomachs should encourage sampling with moderation.)** The stem pith is soft and can be hollowed out for a pipe stem or drinking straw or musical instrument. (The Ojibwa called this plant "flute reed.") Let the stem thoroughly dry before use, as **all parts of the *green* plant except ripe berries and flowers are toxic! Do not confuse with the herbal hemlocks,** which are deadly!

<u>EVENING PRIMROSE</u> (*Oenothera biennis*) – First learn this biennial in its second-year flowering stage and then get to know its first-year appearance when the leaves are arranged in a basal rosette (with no stem). Pinkish pigment in the rosette leaves becomes a helpful identifier. In autumn to early spring, harvest the first-year taproot, peel, and cook to taste. Young stem leaves (in the plant's second year) and flower buds or young flowers can be eaten raw or cooked. **(Don't confuse tough seedpods for flower buds!)** The young succulent stem can be peeled, boiled, and eaten. Mature seeds can be nibbled raw. Dried weed stems make excellent hand drills for fire-kits. (See photo in Chapter 3.)

<u>FERNS</u> – Edibility of ferns must be mentioned here to clear up a misconception that I often hear from my new students, who want to be shown an edible fern called the "fiddlehead." There is no such fern. "Fiddlehead" is a spring stage of young ferns before unfurling to spread their fronds. One fern in the northern half of the East (Ostrich fern, *Pteretis pensylvanica*) is edible in its fiddlehead stage, but a sector of the public apparently believes that all fiddleheads are edible.

Two other ferns are said to be edible in books: bracken and cinnamon ferns. People do cook and eat the young fiddleheads of bracken (see "Bracken" in this section), but I choose not to partake due to the carcinogenic chemicals that eventually develop inside the plant. Cinnamon fern (*Osmunda cinnamomea*) tastes so bitter that I do not harvest it. Its photo is shown for your edification.

<u>**GERANIUM, WILD**</u> (*Geranium maculatum*) – The tannin-rich root is very astringent. Crush and apply topically as a styptic. **Historical**

cinnamon fern (Osmunda cinnamomea)

note: Cherokee warriors carried a pouch of dried, powdered geranium root into battle to sprinkle on wounds, stanch the flow of blood, and stay in the fight.

<u>GINGER</u> (*Asarum canadense* and *Hexastylus arifolia*) – The photograph in Chapter 3 shows two different plants, both called "wild ginger." The uses that follow apply to both. A tea of the root dispels gas and relieves motion sickness. Or one could chew on a tiny section of root comparable to ¼" of packaging string to settle this discomfort. Eating ginger root regularly combats inflammation and lowers cholesterol. Insert pieces of root into meat before cooking. To get children excited about wild plants, simmer roots in a small amount of maple syrup for candy. **People who do not tolerate ginger experience indigestion or diarrhea.** Use in moderation as a test.

<u>GINSENG</u> (*Panax quinquefolia*) – ***Applications to explore through an herbalist:*** Daily intake of root tea works as an adaptogenic (helping the body to acclimatize to new environments and stress). Research shows that this tea (as well as chewing on a dried root) increases mental and physical efficiency. **Hyperactive individuals should avoid using ginseng!**

<u>GRAPE, WILD</u> (*Vitis spp.*) – Grapes are best eaten right off the vine in late summer to avoid the irritation from chemicals that form in juice crushed from the fruit. In spring and early summer a grapevine can be a source of drinkable sweet water. (See Chapter 16.) The vines are very pliable and durable for basketry or for a woven door for a winter survival hut. (See Chapters 18 and 14.) Young shoots, leaves and tendrils are edible, though often bitter. Young leaves showing red/pink pigment usually taste best. Large leaves can be used as a food wrap for baking in a pit oven. ***Other applications to explore through an herbalist:*** The dark-purple, edible grapes contain resveratrol, touted as an anti-cancer and anti-arthritic agent.

<u>GREENBRIER</u> (*Smilax spp.*) – Parallel veins, briers, and thin green tendrils make this common vine easy to identify. In spring the tender new shoots of stem, leaves, tendrils, and briers are delicious raw or steamed. Generally, thin shoots (less than 1/8" thick) and/or purple-pink-tinged shoots are unpleasantly bitter. Look for plump, all-green, new growth for a delicious food. Woody greenbrier vines climbing up trees might seem forbidding with their tough thorns, but carefully pull down the vines to find the succulent spring growth high in the tree (where deer cannot reach). The pulling technique: loosely grasp the thorny vine and, as you begin to pull, slide your hands down the vine and slowly tighten your grip. This bends the sharp points of the thorns so that you can take a firmer painless grip. (Photo in Chapter 3.)

<u>GROUNDNUT</u> (*Apios americana*) – Refer to Chapter 5 for a description of this leguminous vine. Its root produces a "string" of tubers that can be harvested from late fall to early spring. The tubers contain – ounce for ounce – three times the protein of Irish potatoes. Always leave a tuber or two from each vine for future growth. Boil or steam tubers for a delicious meal, but eat while still warm for the best taste. Size of tuber ranges from playing marble to softball (and larger). The mature seeds can be roasted in the pod for twenty minutes then cooked like black-eyed peas. Very young peas, green and succulent, can be eaten after brief boiling.

Groundnut is an important survival food for its availability and protein. Learn the vine intimately, noting the stem's "elbows," so that you can identify it in winter when the leaves are missing. **Nausea, reported by some partakers, is perhaps the result of undercooking!**

<u>HAWTHORN</u> (*Crataegus spp.*) – This spiny, small tree is often found in an old meadow. Its white flowers bloom in spring and early summer. A strip of inner bark cut from a green branch provides ready-to-use cordage. Fruit and very young leaves are edible, but with so many species let taste be your guide. Like its relative – the apple – the seeds do contain some poison, so it is recommended that children not eat the seeds. *Other applications to explore through an herbalist:* The berries, used regularly in a diet (especially when combined synergistically with garlic), lower cholesterol. A leaf paste applied topically to an injury eases pain and swelling through a narcotic action. **Hawthorn's chemicals can be used to alleviate heart ailments but can have adverse effects if used unsupervised! It has been reported that a mere scratch in the eye from one of the tree's thorns can cause blindness!**

<u>HAZELNUT</u> (*Corylus spp.*) – The oil in the leaves (best mixed with rendered fat for application) makes an insect repellent – especially effective against mosquitoes. For quick use (without fat) simply crush leaves into a paste and apply to skin. The sweet nut is edible in the latter part of summer. The inner bark is astringent. By chewing on a twig one can use the astringency to stanch bleeding gums, and so this tree became one of the many "natural toothbrushes" (one twig-end chewed into a "brush") passed down through folklore … like birch, dogwood, maple, etc.

<u>HEAL ALL</u> (*Prunella vulgaris*) – Learn this non-aromatic mint when it is in flower, then learn to identify it by leaves alone before flower heads form, because young leaves are edible and rich in anti-oxidants. Note the bulky structure of the stem's flower head – resembling a small ramrod for a nineteenth-century cannon. *Other applications to explore through an herbalist:* A tea made from the dried flower head contains ursolic acid – a diuretic and anti-tumor compound.

<u>HEMLOCK TREE</u> (*Tsuga canadensis*) – This magnificent evergreen tree contains no poisons like the herbs that bear the same name (water hemlock, poison hemlock). **Whenever the word "hemlock" stands alone in this book, it refers to the tree … not the poisonous herb.** (See description in Chapter 3.) The rarer **Carolina hemlock** (*Tsuga caroliniana*) has a less planar bough with needles angled above and below the branch. Hemlock trees have one of the tiniest tree branchlets found in Appalachia, which makes its lower dead limbs a wonderful kindling source … but only in dry weather. In wet weather or on humid bottomland the persistent bark on these tiny dead twigs can hold too much moisture to be useful to a fire-maker. Hemlock's dead wood makes an excellent fire-making kit. Eat the inner bark raw (sparingly, due to harsh terpenes) or cut it into strips, soak in salt water for 3 hours, rinse, and boil for "pine noodle" pasta. Or dry and grind the inner bark into flour. Its nutrition-list includes: protein, fat, carbs, niacin, thiamine, riboflavin, iron, phosphorous, vitamins A and C. This bark is also antiseptic and a source for pink dye. Needle tea, made any time of year, provides the same nutrients listed above, but should be drunk sparingly. The light-green new growth on

branch tips in spring and summer (1 to 3 growth spurts can occur in a season) is not yet filled with terpenes and edible raw in moderation. Thin rootlets growing at the drip-line can be dug for instant cordage. **Historical note:** Dried, powdered inner bark was applied to the body as a scent-absorber by Cherokee hunters. (See Chapter 4: About Inner Bark and Cambium.)

<u>HICKORY</u> (*Carya spp.*) – Even the smaller branches of hickory have a robust, muscular appearance – never deliquescent. Pinnate compound leaves have 3, 5, 7, or 9 serrate, oblanceolate leaflets. The bark of mature trees may form ridges in diamond shapes. (Other trees with diamond-shaped bark are: locust, walnut, butternut, ash and, to a degree, older tulip trees and sassafras.) Dried leaves mixed with animal fat (bear was preferred by native people) make an insect repellent. Raw crushed leaves smell like *Listerine* mouthwash and repel insects with varying results. Most hickory nuts are edible (those that are not will be obviously bitter or present a medicinal taste), but demanding to extract from convoluted shells. Three hours work might yield a half-cup. The nuts contain proteins, carbs, iron, phosphorous, and trace minerals. Hickory makes a good bow. (The majority of extant Native American museum bows are made of hickory.) The inner bark is an impressive mesh of fibers that can be cut and stripped from a green limb for a tough rope. Even though hickory is one of our toughest hardwoods, dead and dry wood makes an excellent fire-making kit. A spring hickory trunk can be tapped for emergency water (see Chapter 16). There are fifteen species of hickory, and the Eastern U.S. is home to eleven. **Historical note:** An efficient technique for extracting the nutritional material from hickory nuts was employed by Native American women. They crushed shells (not husks) and nuts together and soaked the crumbled mass in water. Within a day or two the nut meat and oil rose to the surface, while porous shell pieces saturated and sank. A fatty soup stock was skimmed off the top. The Algonquins called this "powcohicora," giving us the common name of the tree. The Cherokees called this product "kunuche."

<u>HOG PEANUT</u> (*Amphicarpa bracteata*) – This delicate vine can sprawl on a forest floor to make a fragile ground cover. It climbs tree trunks but not very high. With its three leaflets it might be mistaken for one of the tick trefoils (also a legume) until its fruit develops. It takes good exposure to sunlight for this vine to produce its above-ground edible parts. Clusters of white and/or pale purple flowers hang from the axils and produce pea pods, each holding three edible seeds that are dark and speckled when mature in autumn. Where these seedpods are not available, you might find on more robust plants another offering: a petal-less flower at ground level, which makes a pod-less bean that works its way into the dirt. Successful harvest comes more often in loose soil. Rub the bean between your hands to loosen the "skin" (for discarding) then cook like any pea.

<u>INDIAN CUCUMBER ROOT</u> (*Medeola virginiana*) – Beneath the single (immature plant) or double (mature) whorl of leaves, note the fine spiderweb-like fluff that can be gathered by running thumb and index up the stem. The pure white root is delicious raw. (Harvest only where abundant.) **Historical note:** Native fishermen chewed the root and spit on a fish hook to entice fish to bite.

JACK IN THE PULPIT (*Arisaema atrorubens*) – This plant is included for its unique beauty and for practice in plant ID. If chewed raw or after insufficient drying, the corm (underground swelling) chemically and painfully burns the mouth with calcium oxalate! (See description and photo in Chapter 3.) In late summer and fall a cluster of brilliant red fruits stands after the greenery has wilted. Also called "Indian turnip," Jack's underground corm must be sliced and then thoroughly dried before cooking and eating. **Historical note:** Considering the drying time needed, this corm may not rank as a short-term survival food, but the Cherokee, who named one of their mountain villages "Turnip Town," prized it.

JERUSALEM ARTICHOKE (*Helianthus tuberosus*) – Because this sunflower is difficult to differentiate from others, look for: colonial growth in sunny spots, hairy, slow-tapering stems, two dominant half-veins flaring beside the mid-rib, opposite leaves low on the stem and alternate leaves above, winged petioles, and only slight branching high on the stem. The variably shaped tubers can be cooked in the many ways of a potato. *Other applications to explore through an herbalist:* Tubers contain inulin – beneficial in the diet of diabetics and hypoglycemics and proven to aid in fecal bulk elimination.

JEWELWEED (*Impatiens spp.*) – (For a description see Chapter 3.) The entire plant contains a slimy anti-itch agent that works like cortisone. This mucilage is a stellar cure for rashes caused by poisons oak, ivy, and sumac as well as for chigger and mosquito bites (and any other bite or sting that triggers a localized immune reaction). Jewelweed's juice also neutralizes the sting of nettle, fire ants, and over-exposure to astringents on the skin, but some sensitive-skinned people experience burning from the plant gel. Young spring sprouts, having only just emerged a few inches above ground, can be cooked in two changes of water and eaten. For a positive ID submerge a mature, unhandled leaf into sunlit water and watch the leaf appear to transform into the brilliant, reflective silver of aluminum foil.

JOE PYE WEED (*Eutrochium maculatum*) – This tall whorled plant has an uninterrupted hollow stem and therefore lends itself for use as a drinking straw (when your water source is difficult to get to), as a bellows-like blow tube, and as a snorkel. Before using as the latter, check for insect occupants. *Other applications to explore through an herbalist*: The leaf tea is diuretic.

JUNIPER TREE (*Juniperus virginiana*) – Juniper makes a good fire-kit (especially the light-colored sapwood) and a good bow. Bark can be fluffed as tinder. Though a harder wood than cedar, it bears the cedar "gerbil cage" aroma. Fresh shavings repel insects in survival bedding. Contact with foliage can cause dermatitis for some people. Inner bark provides a reddish to mahogany dye. **Historical note:** A folk-use called for chewing berries to coat the mouth (then spitting out the chewed mass) to ward off contagion when visiting a sick person. *Other applications to explore through an herbalist:* Juniper preparations address urinary and kidney problems.

KNOTWEED, JAPANESE (*Polygonum cuspidatum*) – This invasive shrub has a hollow stem (except at the nodes) and spade-shaped leaves. The stem sections zigzag at the nodes. Young shoots are edible cooked. *Other applications to explore through an herbalist:* Roots contain resveratrol, an anti-inflammatory compound used in treating symptoms of Lyme disease.

KUDZU – A fast-growing, invasive vine (it can smother whole sections of forest) of the pea family with 3 broad leaflets per compound leaf. Making flour from its roots is too labor-intensive for a survival scenario, but very young stem shoots and leaves can be eaten raw or cooked. Mature stem fibers provide impressive cordage.

LAMB'S-QUARTERS (*Chenopodium album*) – Common in garden sites, lamb's-quarters' tender new green parts steam into a tasty, nutritious spinach-like meal. The seeds are packed with nutrients, either cooked or dried and then ground into flour. The plant is high in oxalic acid – to be avoided by those whose bodies tend to manufacture "stones."

LIZARD SKIN – This alga grows in a colony on creek-side rocks and moist stream banks in imbricate fashion, like shingles on a roof. The layered plates and the scaly appearance of individual plates resemble a reptile's skin. Gently wash away sand and dirt clutched by the roots. Its pungent taste makes a nice seasoning for other greens.

LOCUST, BLACK (*Robinia pseudoacacia*) – Once called "yellow locust" (wood is golden-yellow), this native Appalachian tree was the Cherokees' preferred local bow wood. It was also used for dart shafts. Locust bark is diamond-shaped. The wood is rot-resistant, making it an excellent choice as a shelter post or stake driven into the earth. Dead trees in moist valleys may provide excellent inner bark fibers for rope or tinder. Unlike the **honey locust** with <u>branched</u> antler-like thorns, the black locust has <u>thorns in pairs</u>, like the horns of a goat, located on the branches at the base of leaf stalks. While some foragers eat the fresh petals and cooked young beans, I have not eaten from this tree so I cannot advise. **Some researchers consider all parts of black locust poisonous!** Also, on this tree look for a horizontal (shelf-like) bracket fungus growing on the outer bark. This fungus, cracked-cap polypore (*Phellinus rimosus*), if dry, can be used as a coal-extender or transporter. Cut a section from the flaky underside and let a hot coal smolder through it as you travel to a new camp. The woody topside can be used as a hearth for a fire-kit with, among other woods, a spindle of white pine.

LOCUST, HONEY (*Gleditsia triacanthos*) – While black locust has once-compound leaves, honey locust has once- or twice-compound leaves. Its thorns are <u>branched</u> and scattered along the trunk. With a little pruning and carving, thorns can be adapted to fishhooks. Honey locust's twisted bean pods are longer than black locust pods and contain a sweet edible pulp before ripening. **Black locust pulp may be toxic! Learn to distinguish one from the other!**

MAPLE (*Acer spp.*) – The inner bark can be eaten raw, boiled, or dried and crushed to flour. Refer to Chapter 4: *About Inner Bark and Cambium.* Inner bark tea (plus crushed leaves) can be strained and used as a topical wash on serious burns, fighting infection by sealing the wound from air, relieving pain, and promoting regeneration of tissue. Roasted or boiled nuts (sheath and wing removed) are edible. Maple wood bends well (without heating) for use in a wickiup frame. Maples make good fire-kits. Box elder shoots grow fast and straight and make excellent hand drills. All maples can be tapped in early spring for emergency potable water. (See Chapter 16.) (Boiling down to syrup is not a practical survival project.)

MAYAPPLE (*Podophyllum peltatum*) – Fruit will appear on forked stems near the fork, but never on unforked stems. **The fruit is edible when ripe, poisonous when unripe!** The green fruit turns yellowish-brown and, when mature, falls off the stalk when handled. **Historical note:** Cherokees crushed this highly toxic plant and steeped it in water to sprinkle on potato plants to ward off the potato beetle. **Avoid plant juice contact with skin or eyes!**

MEADOW BEAUTY (*Rhexia spp.*) – Note the flower's unique yellow anthers contrasted against the magenta petals. Young lemony leaves can be eaten raw or cooked. One or two tubers can be found on a single plant. Dig them up, clean, and eat raw or cooked.

MEDICK, BLACK (*Medicago lupulina*) – To differentiate from hop clover, look for a tiny bristle tipping each medick leaflet. Gently roast the seeds, then eat or crush to flour.

MILKWEED, COMMON (*Asclepias syriaca*) – For a delicious, asparagus-like dish, harvest the supple spring shoots (just emerged from the ground with leaves still pointed skyward) and drop stems and uppermost leaves into boiling water until tender. (The tender tops of slightly older plants may be used if they snap off.) Later in the summer boil the immature flower buds in the same manner. Still later, the <u>very young</u> seed pods (under 1"-long, firm) can be boiled and eaten whole. Slightly longer (but still immature) pods can be cooked for their fleshy white edible interior. **Do not confuse dogbanes (*Apocynum spp.*, all poisonous) or any bitter species of milkweed for common milkweed! Learn to differentiate these plants with certainty!** Stem fibers of mature plants make strong cordage.

MINTS (*Mentha spp.*) – All mints have square stem cross-sections. Used raw, most mints make fine mouth fresheners, nibbles, and spices. Leaf tea stimulates the digestive system with moderate laxative action. This same tea (drunk or used topically) works as a refrigerant to cool the body. (Photo in Chapter 4.)

MULBERRY (*Morus rubra*) – The scratchy leaves of native red mulberry trees have the same three varying shapes of sassafras leaves, but mulberry's margins are serrate. Red mulberry has purple-black mature fruit. **Only ripe fruit should be eaten!** Introduced white mulberry (*M. alba*) has un-lobed, smoother-textured leaves and white-pink, less tasty, mature fruit. The green inner bark fibers of both species make excellent cordage. Used topically, sap is anti-fungal. **Except for <u>ripe</u> fruit, all parts of mulberry are considered toxic!**

MULLEIN (*Verbascum thapsus*) – An introduced Mediterranean plant, mullein makes a basal rosette of fuzzy, cushiony, pale-green leaves in its first year and in its second year produces a flower stalk (with leaves). Leaves add insulation inside winter shoes and clothing. The dry winter flower stalk has many open seed capsules for dipping into melted tallow for a torch. The crushed seeds are a fish poison (see "Buckeye"). A straight dry winter stalk can be smoothed of rough spots and spun as a fire-maker's hand drill. Sun-dried leaves make a nice addition to tinder. ***Other applications to explore through an herbalist:*** For bronchitis (or as a safeguard when entering pollen season), the mucilage in the green leaves can be utilized for anti-inflammatory and antibiotic applications. Flower tea can be brewed for stomach virus. Flowers soaked in olive oil can be used as eardrops to fight infection.

<u>NETTLE, STINGING</u> (*Urtica dioica*) and <u>WOOD</u> (*Laportea canadensis*) – With opposite and alternate leaves respectively, stinging and wood nettles both possess tiny hairs that prick the skin and, pumping like a hypodermic needle, inject formic acid (the same chemical injected by fire ants) from a sac. (See **curly dock** and **jewelweed** and **plantain** for antidotes.) Drop the young plant or the new growth from an older plant into boiling water briefly to convert acids to protein, wilt the hairs, and render the greens edible. Extremely nutritious, iron and protein-rich. [False nettle (*Boehmerica cylindrical*) without stinging hairs is similarly edible.] Mature stem fibers make strong cordage. ***Other applications to explore through an herbalist:*** Deliberately stinging oneself drives uric acid from joints, alleviating the discomfort of arthritis and rheumatism.

<u>NEW JERSEY TEA</u> (*Ceanothus americanus*) – This small shrub's leaves provide a pleasant caffeine-free tea. The root's core contains tannin (astringent) and saponin (natural soap). ***Other applications to explore through an herbalist:*** Dried root-bark gathered in summer can be prepared as an anti-histaminic, hypotensive tea that lowers blood pressure and sugar levels.

<u>OAKS</u> (*Quercus spp.*) – (See Chapter 3, *A Closer Look at Some Old Friends.*) It takes two years for a red oak acorn to mature … one for a white oak acorn. This generally renders the red acorn more tannic (bitterly astringent) to taste. Since the tannin must be leached out before ingestion, by choosing white oak acorns in the fall you have an easier job of preparation, though – once the leaching is done – some people prefer the taste of red oak acorns. White oak acorns root late in the same fall in which they fell; red oak acorns lie dormant through winter and root the following spring, making red nuts found in winter more nutritious than whites.

Many people speak of eating acorns raw, and though I have occasionally tasted some palatable, unleached white oak nuts (especially the coastal live oak), the chance for upset stomach looms in the future for such a forager. (For acorns as food see Chapter 18, *A Survival Meal in Autumn*). Like all native nuts, acorns contain saturated fats, serotonin, protein, and no cholesterol. Chestnut oak (a white oak) of the mountains has a bark that resembles red oak and makes one of Appalachia's largest native acorns. (Growing in the limestone country of Alabama, Tennessee, and Kentucky, the bur oak makes the largest.) Oak wood can be used to make a bow.

<u>PARTRIDGEBERRY</u> (*Mitchella repens*) – This low, evergreen, trailing vine with dark-green, opposite leaves makes a red fruit through the joint effort of two flowers. Besides the stalk scar (after being picked) the fruit bears two dimples where the flowers once connected to the ovary. These dimples are visible in the photo on page 125, which shows partridgeberry growing on moss. Though many books describe this edible berry as "insipid," a subtle apple taste is present.

<u>PAWPAW</u> (*Asimina triloba*) – Oblanceolate leaves alternate by 180° intervals, giving this tree's boughs a planar appearance. The rubbed leaves smell like bell pepper or tomato vine. This aroma is insecticidal, useful as an insect repellent. After testing for tolerance, apply crushed leaves to the skin and clothing. As a chigger guard, bruised leaves can be stuffed inside clothing or laid down for a "survival bed" in summer. The fruit is sweet (a combination mango/banana/ pineapple), but **some**

people contract dermatitis or gastrointestinal upset from handling and eating it. **Sample in moderation!** The fruit's seeds can be dried, crushed into powder, and applied to the scalp to kill lice. Green inner bark can be cut into strips for cordage. The dead wood is excellent for a fire-kit, especially as a hearth for a hand drill.

PEPPERBUSH, SWEET (*Clethra alnifolia*) – This tall shrub with spare numbers of leaves grows in moist areas, especially favoring shaded mountain stream banks among rhododendrons. Its cinnamon bark peels in small, striated shards or patches revealing a smooth trunk beneath. The leaves contain saponin, a natural soap that lathers in the hands. (Photo in Chapter 4.)

PERSIMMON (*Diospyros virginiana*) – This tree flourishes at the edge of (or in) a meadow though occasionally a deep-forest, female tree can be found bearing fruit. The mature tree is checkered with squarish, chunky blocks of thick gray bark. The autumn fruit of this tree is ready to be eaten raw when it is mushy. Premature fruit is highly astringent and will provide an unforgettable experience of "cotton-mouth" if chewed. Smear the deseeded fruit pulp on an oiled, barkless log and dry by a fire for fruit leather. Leaf tea is high in vitamin C. Remove the membranous sheath surrounding each seed by vigorous buffing with a handful of grass, then roast the seeds by a fire (or on a tray at low heat in your oven with the oven door ajar) until a delicious rich caramel aroma fills the air. Grind the seeds and then brew for an unforgettable hot cocoa substitute. As a member of the ebony family, the wood is hard and durable as a tool.

PINE (*Pinus spp.*) – All the true pines of the East have edible inner bark, but raw bark should be eaten sparingly due to harsh terpenes. (Refer to Chapter 4: *About Inner Bark and Cambium.*) Nutrients include protein, fat, carbohydrates, phosphorous, iron, thiamin, riboflavin, niacin and vitamins A and C. To dispel terpenes, cut into thin strips, soak in salt water for several hours, wash, and boil as "pine-noodle pasta." Inner bark can also be dried and ground into flour. Gently fry strips of inner bark in oil to make "pine bacon": Heat both sides of a flat stone on hot coals and oil with animal fat rendered from the flesh side of a skin like raccoon or beaver. The older the pine, the tastier the bark.

Pine seeds harvested from female cones and extracted from their sheaths are edible. Gather from closed cones that have lost most of their greenness. Young male cones (pollen-makers) can be boiled and eaten. When mature these male cones provide edible pollen. Young green branch shoots and needles are edible in moderation with a little care taken to remove the brown, papery fascicles where the needles bundle together. Tea can be brewed from mature needles at any time of year and, ounce for ounce, will provide five times the vitamin C of lemons, but the terpene level suggests using only sparingly.

Globs of semi-hardened sap can be found at trunk wounds. Climb for these globs or pry them loose with a long stick. Raw sap can be used topically as an antiseptic on cuts. Heated sap (as hot as can be tolerated) can be placed on the skin to draw out splinters as the sap cools. For glue, melt this sap by a fire (careful, it is highly flammable) on the concave side of a rock and then sprinkle into it the ashes from burned hardwoods. Mix well and you have a superb epoxy glue that hardens

like a rock as it cools. Start with 5 to 1 ratio of sap-to-ash. Desired proportions can be learned by experimenting. Too little ash makes flexible glue (sometimes desirable). Too much ash makes brittle glue (seldom desirable). Dig up pine rootlets from the canopy drip-line for ready-to-use rope. The most easily acquired tinder in the present-day East is available in November when "rolling clouds" of pine needles accumulate on urban roads to be ground into fine fibers beneath automobile tires. The pines known to make fire by friction are white, short-leaf, hemlock, and fir.

PINE, VIRGINIA (*Pinus virginiana*) – This pine with short (3" or less), twisting needles (in 2's) is rich in a sap that intensifies its flammability after the tree dies. This is the famous mountain "lighter-wood" – a superb kindling. When the tree dies and falls, its resin dehydrates and condenses over time <u>at the core of the tree</u>. Because the resin is antiseptic, this tree rots from the outside in. The remaining "spine" of the

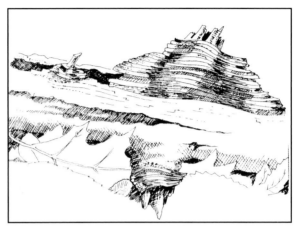

dead Virginia pine

tree contains glassy amber deposits running in "veins" through the wood, occasionally almost as hard as petrified wood. Cut slivers of this resin-rich wood for volatile kindling.

PINE, WHITE (*Pinus strobus*) – (See Pine above.) White pine leaf bundles are comprised of 5 needles, and limbs are arranged in whorls. This is one of few pines that can be reliably used to make a fire-kit. As a food, its inner bark arguably surpasses all other pines – especially if the tree is an old one. ***Other applications to explore through an herbalist:*** Inner bark tea serves as an expectorant. The pine's antiseptic mucilage is soothing to the mucosa.

PIPSISSEWA (*Chimaphila maculata*) – Two plants bear this common name. The more common pipsissewa of southern Appalachia is also called "striped" or "spotted wintergreen." The native name means "breaks apart the stone." (Description and photo in Chapter 3.) ***Applications to explore through an herbalist:*** A leaf tea (drunk in moderation to dissolve stones) contains arbutin, ursolic acid, and sitosterol – chemicals that are diuretic, antibiotic, and antiseptic to urinary tracts.

PLANTAIN, BROAD-LEAVED (*Plantago major*) – (See Chapters 3 and 5 for photo and description.) A mature leaf can be crushed/chewed and applied topically to neutralize venom from an insect sting and to soothe poison ivy rash and nettle sting. A poultice of crushed, soaked leaves can draw out infection from a wound, promote tissue regeneration (by its allantoin content), and ease swollen hemorrhoids by its anti-inflammatory properties. It is also antiseptic. To build up insect repellency for the summer, eat daily the tender seeds (the top inch) clustered on a spike that rises from the center of the basal leaves. These seeds contain psyllium, a fiber noted for intestinal cleansing and lowering cholesterol. ***Other applications***

to explore through an herbalist: **Narrow leaved plantain** (*Plantago lanceolata*, also called "English plantain") has longer lanceolate leaves and a taller flower stalk with a bullet-like head at top from which tiny greenish-white flowers emerge. Its leaf tea lowers cholesterol and decreases blood pressure.

<u>PLANTAIN, PALE INDIAN</u> (*Cacalia atriplicifolia*) – Growing over 6' tall this plant produces flat clusters of small white flowers in late summer. Its stem and underside of leaves are covered with a white fungal bloom (looks like powder). Crushed leaves can be used as a poultice for insect stings to draw venom from just under the skin. (Plantains cannot be expected to draw venom from intramuscular injections like snakebite.)

<u>POISON IVY</u> (*Toxicodendron radicans*) – (See Chapter 10.)

<u>POKEWEED</u> (*Phytolacca americana*) – The folk name "poke salad" implies that the plant can be eaten raw. Researchers report that the **raw plant and berries are toxic! Concentrated plant juice may cause dermatitis and damage chromosomes! Respiratory failure (and death) has been attributed to ingesting raw pokeweed!** The purple pigment is the plant's poison. To enjoy the <u>cooked</u> green, harvest young leaves still angled skyward, boil in two changes of water, and eat as a delicious sallet. See photo in Chapter 3. **Historical note:** "*Sallet*" is a Middle English word meaning "cooked green," hence the corruption to "poke salad."

<u>POPLAR</u> (*Populus spp.*) – The inner bark of all poplars (like cottonwood and aspen) is edible raw or cooked. See Chapter 4: *About Inner Bark and Cambium.* Dead wood is good for a fire- kit. (Note that the tree commonly called "tulip poplar" or "yellow poplar" is <u>not</u> a poplar.)

<u>PRICKLY PEAR</u> (*Opuntia humifusa*) – This cactus is often found around sandy rock outcrops. The pulp of the fruit (skinned and deseeded) is edible as is the leaf pad. Both should be carefully cleaned of bristly tufts of spines and rind. Eat the pad raw or slice and cook. The seeds can be dried and then crushed into flour. Flower petals are edible. **Avoid skin contact with both thorny and hair-like spines!**

<u>PURSLANE</u> (*Portulaca oleracea*) – Often found in old garden sites, purslane has very succulent leaves growing on a fleshy stem that might sprawl on the ground or grow erect. Note the reddish tint to stem and branches. Eat delicious leaf and stem shoots raw or cooked. Contains iron, calcium, phosphorous, vitamins A and C.

<u>QUEEN ANNE'S LACE</u> or **wild carrot** (*Daucus carota*) – **It is essential to know how to distinguish wild carrot from poisonous hemlock herbs, as they produce similar flower arrangements! One factor is that Queen Anne's lace has hairs on stems and stalks, while the hemlock herb does not. Rather than the carroty smell of Queen Anne's crushed leaves, poison hemlock's leaves emit an unpleasant odor. Leaves of water hemlock, also deadly, do smell carroty!** There are two wild carrot photos in the color section. One shows a second-year plant with flower and a first-year basal rosette in the foreground. The other is a leaf close-up. Dig up the pale, first-year root in fall or winter after its first growing season, core it, peel it by lightly scraping, and eat raw or cooked. **Note: Poison hemlock's deadly root looks like a white carrot!** Queen Anne root can be crushed, mixed with lard, and

applied to burns as a healing agent. The very young tender flower stalk is edible early in the plant's second season. Discard leaves and rind and eat raw or cooked.

<u>REDBUD</u> (*Cercis canadensis*) – The heart-shaped leaves have the feel of balloon rubber. Before leaves open in spring, enjoy eating the pink-magenta flowers that cluster along the limbs. The young, tender pods and peas that develop from the flowers can be steamed, lightly boiled, or sautéed in butter for a cooked vegetable. The dead wood can be used for a fire-kit.

<u>RIVERCANE</u> (*Arundinaria gigantean*) – Our native cane, once so important to the Cherokees in crafts (blowgun, basketry, flute, fish trap, arrows, spears, etc.) comes in two sizes – the smaller, sparsely-growing "switch cane" and the taller variety found in thick brakes on floodplains. Easily mistaken for bamboo. Cane bow/arrow-making is covered in *Secrets of The Forest, Volume 4*. Dead cane can be scraped to make dry tinder in wet weather. A friction fire-kit of cane is covered in *Secrets Of The Forest, Vol. 2*. **Rivercane contains precursors to cyanide!**

<u>SASSAFRAS</u> (*Sassafras albidum*) – This medium-size tree of the South can grow to a giant in ideal conditions (I found one in alkaline soil next to the Buffalo River in Arkansas with a trunk diameter of 5'.) Sassafras has three (or more) classic leaf shapes: football, mitten, and "Casper the Friendly Ghost" – all with smooth margins. (Oldest trees seem to default to elliptical leaves only.) Scraped outer bark reveals a cinnamon color underneath. Rubbed leaves smell like *Froot Loops* breakfast cereal, the roots like root beer. The young leaves and leaf buds are edible. Mature leaves gathered in summer can be dried and then crumbled into soups as a thickener. **The root was the target of the Surgeon General's research done decades ago and was proclaimed to be carcinogenic and to cause liver damage. The culprit compound, safrole, was injected into laboratory rats until tumors developed. This test has been criticized as flawed. (To put things into a proper perspective, a can of beer is fourteen times more carcinogenic than a cup of sassafras tea, yet the root remains black-listed. One must decide for himself about the prudence of ingesting sassafras!)** For centuries southern Appalachian folk have drunk this root tea as a spring tonic to thin the blood and to prepare one for the hot months of summer. Researchers have shown that it boosts the immune system, reducing the chances for sore throat and congestion during the spring allergy season. This is a tea I drink when nauseated from having consumed bad water or spoiled meat, because it kills the problematic microorganisms (good ones, too, unfortunately) introduced into the gut. The root bark can be pounded and inserted into meat as a spice. The dead, thoroughly dry wood makes a fire-kit, an important entry on the fire-making list because it can grow on higher, dryer ground compared to most fire-making trees, which live in low, humid areas, where dead wood is often damp. Wood shavings can be spread on the ground beneath a primitive bed as a barrier against insects and mites. ***Other applications to explore through an herbalist:*** Sassafras root tea can be mixed with yellowroot tea (in specific proportions) for an adaptogenic (see ginseng). **Historical note:** "Sassafras" is a native word that translates as "green twig" – a feature that might aid in winter identification. (Photo in Chapter 5.)

SERVICEBERRY (*Amelanchier arborea*) – There are varieties of this small tree that develop edible purple-black fruit, but in the Southern Appalachians I find most mature fruit to be red to dark red and, oddly, exclusively fruiting near a waterfall. Elliptical leaves are 2"-3 ½" long with quite straight lateral (pinnate) veins, fine orderly teeth, and a slightly cordate base. The spidery white flowers appear early in spring, brilliant against the bare, wintry guise of the forest.

SILVERBELL (*Halesia carolina*) – When young, this tree's bark shows handsome streaks of white and blue-gray. Much older, the gray bark takes on a bluish cast, like armor. The leaves hide a silky thread of latex inside the mid-rib vein. In spring, fruit forms with 4 wings running longitudinally. In mid-spring, before these fruits harden to nutlets, eat the tender green fruit entire.

SOCHANI (*Rudbeckia laciniata*) – Also known as "green-headed coneflower", this creek-side plant remains a favorite cooked green of Cherokees. Gather the lighter-green new growth off the tops of plants and steam or boil as a potherb until palatable. My students often misidentify sochani, as it has variable appearances, and so I recommend getting ID verifications from those who know the plant. **Do not confuse with cowbane or water hemlock – both deadly!**

SOLOMON'S SEAL (*Polygonatum spp.*) – Flowers (in late spring) and later fruit (summer) hang down from the leaf axils. These can be solitary flowers (and later berries) per axil or grouped in twos or threes (or more) from forked stalks. To distinguish this from false Solomon's seal when flowers are not present (false Solomon's seal has a plume of flowers that are terminal on the stem), check the underside of a true Solomon's seal leaf for a whitish powder – like the fungal bloom wiped off a plum. (False Solomon's seal lacks this.) Solomon's seal's rhizome is a delicacy. Eat it raw or cook it as you would a potato. The very young supple shoot (remove the closed leaves from the stem) is also a candidate for a cooked green. The berry flesh is edible with an interesting nutty taste. Please harvest the plant sparingly, leaving behind parts of rootstocks to ensure a long-term population of these lilies. ***Other applications to explore through an herbalist:*** The root starch contains inulin, recommended for people suffering from blood sugar disorders. **Historical note:** The root accumulates crater-like scars where former stems once attached. This relief pattern gave the plant its name by resembling a cluster of royal seals that were pressed into sealing wax.

SORREL, WOOD (*Oxalis spp.*) – Often confused with clover, sorrel has three distinctly heart-shaped palmate leaflets that connect at their points. Sorrel leaves, flowers, and fruit (which looks like a tiny okra) can be nibbled raw. The lemony taste of sorrel can help ease nausea and dry heaves. Simply crush a handful of green leaves in a glass of cold water and sip. **Large amounts of sorrel are not recommended due to oxalic acid content, which bonds with free calcium in the body to form stones (crystals of calcium oxalate) in the urinary tract! Historical note:** Oxalic acid is present in many wild foods. The Cherokees were prone to stone formations, which is why the legacy of their plant medicines includes a number of stone remedies.

SOURWOOD (*Oxydendrum arboreum*) – This contortionist tree can make quite a serpentine climb up into the forest canopy, yet some of its vertical branches and survival shoots from the root are perfectly straight and prime material for short-range (it's a dense, heavy wood) arrow shafts. The bark of older trees is coarse with pronounced ridges, whose cross-sections have angled sides like a gold ingot. Note very fine teeth on the leaf margin and tiny hairs that follow the veins on the underside of the leaf. **Historical note:** Though the leaf was not considered a food, Cherokees chewed it for its tart juice when drinking-water was not at hand. Oxalic acid triggers saliva glands to flood the mouth for diluting the acid. The result is pseudo-thirst-relief.

SPARKLEBERRY (*Vaccinium arboreum*) – This crooked understory tree produces sweet blue-black berries in early fall that stand on he tree into early winter. Note the trunk's shredded bark and the blunt-tipped, dark, glossy leaves. Green leaves persist for a time in winter.

SPICEBUSH (*Lindera benzoin*) – Just brushing past this shrub releases its aromatic essence into the air (an airborne sign to a tracker). The smell of a bruised leaf or twig is reminiscent of grapefruit rind. The fruit – fire engine red – can be nibbled but mostly it is used as a substitute for allspice in cooking. (Split, deseed, and dry for two weeks, then grind into a powder.) Insert green twigs into wild meat to tame a "gamy" taste. **Other applications to explore through an herbalist:** A green twig tea may be useful, topically or ingested, to fight yeast infection.

SPIDERWORT (*Tradescantia virginiana*) – Look for long, robust, grass-like leaves extending from the stem and blue to purple, 3-petaled flowers with 2 smaller modified leaves (bracts) beneath each flower. Young, tender, green shoots (leaves, stem) are edible raw or steamed.

SPRING BEAUTY (*Claytonia spp.*) – Swollen roots can be cooked like a potato and eaten. Look for large individual swellings that give rise to multiple stems. White flower petals are pink-veined. One pair of leaves usually grows halfway up each stem. Harvest greens and flower when they emerge in spring (eat raw or cooked) or the root late in its season after the plant has wilted.

STRAWBERRY (*Fragaria virginiana*) – The nutritional value of wild strawberries is higher than cultivated varieties. *Common strawberry*, (photo on page 132) with rounded leaf tips, offers a leaf tea (to be drunk) to resolve diarrhea. The pointed leaves of *wood strawberry (Fragaria vesca)* make a leaf tea wash (used topically) for sunburn. [Indian or false strawberry *(Potentilla indica)* has yellow flowers and protruding red seeds on the fruit surface; and though its taste is subtle the fruit is edible, as are the cooked young leaves.] **Other applications to explore through an herbalist:** Fruit and dark seeds contain ellagic acid, a cancer preventative and hemostat for bleeding wounds. **Historical note:** Cherokees held the fruit pulp on their teeth to dissolve tartar.

SUMAC – Of the common sumacs – **winged** (*Rhus copallina*), **smooth** (*R. glabra*), and **staghorn**, (*R. typhina*) – Handsome, golden-yellow, dead wood makes a usable bow-drill fire-kit, but the large soft pith forces modifications. For example, if a whole limb is used as the drill, the points on either end of the drill must be tapered off to one side rather than being centered. Such a wobbly drill requires

extra strength from the fire-maker and would not be recommended for a beginner. The tannin-rich root of smooth and staghorn sumacs can be made into a gargle for sore throat by steeping a section of scored root (¼" wide, ½" long) in one cup of hot water. In spring, harvest new succulent shoots, carve away the thin green rind and eat the shoot raw for a fruity snack. In late summer the red berries of smooth, staghorn, and winged sumacs contain tasty malic and ascorbic acids that can be leached into a pot of cool water. (Also present is bitter tannic acid, which dissolves more slowly in cold water.) Harvest the berry clusters when deep red. Before submerging berries in cool water remove all critters, spider silk, and frass (caterpillar droppings). When the water turns pink, strain the liquid through a cloth to remove tiny hairs that would otherwise irritate the throat. Sweeten (or not) and drink or freeze for "sumac-sicles." **Historical note:** The Cherokees called this drink "qualla" – the same name they gave to their North Carolina homeland. (See Chapter 3 for descriptions and photo.) **Poison sumac (*Toxicodendron vernix*) berries grow from leaf axils and mature to a grayish-white hue!**

<u>SWEETGUM</u> (*Liquidamber styraciflua*) – Able to repair its damaged inner bark, sweetgum tolerates cutting into its bark to release antiseptic sap, which can be rubbed into a dirty wound or chewed to ooze over a sore throat. Dead wood can be used to make a fire-kit, but because sleeves of bark are resistant to rot the interior wood is often damp and will need drying. **Historical note:** Native Americans likely added dried fruit to this "chewing gum" medicine to make it tolerable.

<u>SWEETLEAF</u> (*Symplocos tinctoria*) – This semi-evergreen shrub often forms whorled branches. The trunk's gray bark is very clean and remains so even when reaching tree-size. <u>The leaves are white-hairy on the underside.</u> Though not considered an edible, the leaf can be chewed in mid-summer through early winter for its sweet-and-tart juice. (See tasting technique and photo in Chapter 2.) **Do not confuse with mountain laurel, rhododendron, and azalea (all toxic)!**

<u>THISTLE</u> (*Cirsium spp.*) – The various thistles seem unlikely foods because of their formidable spines, but trim the young leaves of these prickles and eat raw or steamed. Strip an older leaf of its blade, leaving the edible mid-rib vein and stalk (which you should peel by scraping before eating). Enjoy a mild wintergreen taste in raw first-year roots. (To soften older roots, boil and scrape away outer rind. The young stem of the second-year plant is known as "survival celery." Scrape away its rind and eat raw. (Photo in Chapter 3.) **Historical note:** Thistle down from mature vase-like seed pods was used to fletch darts of Cherokee blowguns.

<u>TULIP MAGNOLIA</u> (*Liriodendron tulipifera*) – While other native tree's leaves terminate in a sharp or rounded point, the tulip tree leaf ends in a broad vee-notch. In spring the bark can be cut off in sheets for roofing material for a shelter. This same bark can be shaped into containers. (See Chapter 5, *Making a Bark Basket.*) Dead wood makes a fire-kit (often mediocre because the porous wood can hold undetectable moisture). Yet it dries fast and burns quickly in the first stages of a pyre. Dead inner bark fibers are easily found for cordage and for tinder. **Tulip tree is not a poplar but a magnolia and offers no raw edible bark! Historical note:** The Cherokees made their dugout canoe from a felled tulip tree.

<u>VIOLET</u> (*Viola spp.*) – Leaves and flowers (**but not seeds!**) of **common** (*V. papilionacea*), **wood** (*V. palmata*), **bird's foot violets** (*V. pedata*), and a few other species are edible raw or steamed. ***Other applications to explore through an herbalist:*** Leaves contain salicylic acid (an anti-inflammatory) and rutin (a mucilage that makes capillaries more elastic to prevent bruising). (See photo in Chapter 3.) **Yellow violets are not edible!**

<u>WALNUT, BLACK</u> (*Juglans nigra*) – The nutritious nut is delicious and easily extracted once the shell has been correctly cracked (see Chapter 7: *Nut Cracking*). The green to black hulls can be soaked in water for a rich, dark brown dye. Husks can be used as a fish poison (see Buckeye). **Do not expose non-callused skin to quantities of the husk's dark pigment. Astringency causes severe stinging! (This can be alleviated by application of jewelweed.)** Green or dead inner bark fibers make cordage. The wood is extremely strong and durable. **Historical note:** In deep woods, these trees often mark the locations of former open spaces where pioneers homesteaded.

<u>WILLOW</u> (*Salix spp.*) – Inner bark is anti-inflammatory and contains salicin, a precursor to aspirin, but takes 6-8 hours to be metabolized. (See Chapter 5: *Easing a Headache*.) Boil inner bark for an emergency food. Dead wood makes a good fire-kit; but, because willows grow in wet areas, drying the wood may be necessary. Dead inner bark makes good tinder and cordage, depending on the state of decomposition. (Photo in Chapter 3.)

<u>YARROW</u> (*Achillea millefolium*) – Crushed, fresh leaves stanch the flow of blood from wounds and impart a pleasant fragrance to the body. ***Other applications to explore through an herbalist:*** The occasional use of yarrow leaf tea is helpful as an expectorant, analgesic, and sweat-inducer. **Use only under professional guidance. Overuse may damage the liver!**

<u>YELLOWROOT</u> (*Xanthorhiza simplicissima*) – This creek-side plant contains berberine, the same healing compound found in goldenseal. Berberine is soothing and healing to burned tissue and to irritated mucosa. For a mouth/lip sore chew a small piece of dried or fresh root (comparable to a quarter-inch of packaging string) and hold it against the sore. Berberine is astringent and kills sore-causing bacteria. Swallow the bitter juice to relieve nausea; spit out pulp. ***Other applications to explore through an herbalist:*** Berberine stimulates the immune system, drops blood pressure, reduces inflammation, and serves as an adaptogenic when mixed with sassafras root tea. (See description in Chapter 3.) **Some researchers warn that large dosages or frequent use of yellowroot can be toxic! Others advocate frequent use.**

<u>YUCCA</u> (*Yucca filamentosa*) – The root and leaf can be pounded for its natural soap – saponin. Inside yucca leaves one can find strong fibers for cordage. (See Chapter 13 for directions.) Summer flower blossoms are edible raw. Young flower stalks can be peeled, boiled, and eaten. The winter-dried flower stalk serves as a hand drill in a fire-kit. (Harvest these stalks when green just after flowering before insects can invade with tunnels that weaken the stalk. Then lash stalks to a straight support to dry under a shelter where campfire smoke will discourage infestation.) **Historical note:** The root juice was used by Cherokees as fish poison and dart/arrow poison for small prey. Manioc root (or cassava) is also called "yuca" and sometimes misspelled by grocers as "yucca." Yucca and yuca are two entirely different plants!

agrimony

air potato

alumroot

amaranth

ash

autumn olive

bamboo

basswood

bear paw rosinweed

beautyberry

beech

blackberry

bloodroot

highbush blueberry

buckeye

bugleweed

bulrush

chufa

cattails

black cherry

black cohosh

corn salad

ox-eye daisy

dogbane

flowering dogwood

duck potato

elderberry

evening primrose

geranium

ginseng

grape

groundnut

hawthorne

hazelnut

heal all

hemlock

hickory

hog peanut

Japanese knotweed

Indian cucumber

Jerusalem artichoke flower

Jerusalem artichoke

jewelweed

joe pye weed

juniper

kudzu

lamb's quarters

lizard skin

black locust

black locust with polypore fungus

honey locust

box elder maple

red maple

mayapple

meadow beauty

black medick

common milkweed

mulberry (red above, white below)

mullein

wood nettle

New Jersey tea

red oak

white oak

partridgeberry

pawpaw

persimmon

Virginia pine, dead

white pine

pale Indian plantain

pokeweed

poplar

prickly pear

purslane

Queen Anne's Lace

Queen Anne's leaf

quickweed

redbud

rivercane

serviceberry in early spring

serviceberry in late spring

silverbell

sochani

Solomon's Seal

sorrel

sparkleberry

sourwood

spicebush

spiderwort

spring beauty

common strawberry

sweetgum

tulip tree

walnut

yarrow

yellowroot

yucca

"The patrons stroll down the aisles of the supermarket, and I think of my days in the Pleasant Valley and each spot where I had collected my food … leaves of sochani by the shoal, tubers of spring beauty near the old chestnut log, and crayfish in the bathing pool."

~ Stoney St. Ney, The Last Real Place

CHAPTER 7
Browsing the Green Market
~ Edible Plants of the Wild ~

Listed below as a quick reference are the parts of wild plants that are edible. Most are covered in previous chapters of this book. Others are offered here as a compilation for your future research. Refer to the color photo section and to Chapters 2-6 for details on specific plants.

CATEGORIES OF PLANT PARTS AND COMMON EDIBLE EXAMPLES

Inner Bark and/or Cambium

Read the section in Chapter 4: *About Inner Bark and Cambium*. Trees with edible inner bark are: **all native pines of the Eastern U.S. (including hemlock, spruce, and fir), maple, willow, beech, basswood, slippery elm, true poplars, and birches (in moderation).** This survival food is available generally in spring, because inner bark gradually hardens and becomes bitter through the growing season. But the evergreens on this list can often provide some sustenance outside the spring parameter, especially if cooked.

Extract inner bark from a limb (not the trunk) by carefully shaving away and discarding outer bark. Pry up the remaining inner bark and peel it away from the underlying sapwood, which is a smooth, glossy sheen of harder cellulose (not edible). The inner bark can be eaten raw, but the terpene content in pines calls for moderation to prevent stomach irritation. To reduce terpene content, solid pine inner bark can be boiled into "noodles" or fried into "bacon." Dried and powdered into flour, it can be used for baking or soup-thickening. Maple and willow are more palatable if boiled. If a ribbon of inner bark is decidedly bitter even in spring, use the preferred slushy cambium.

Each year the inner bark contributes in two directions to the growth of the tree – thickening the outer bark as well as the inner sapwood. (This is how growth rings form and outer bark gets crustier.) Expect to find hardened inner bark in late summer through winter in most trees.

Roots and Tubers

Some of the best foods of plants are to be found as underground swellings packed with nutrition. During the growing season tubers store up nutritional material for the next spring's growth, so harvest them from late fall to early spring. Dig with meticulous care and patience, using the shovel-end of a digging stick. Use the blunt end of the stick for applying pressure from body weight leaning into the stick. To be certain that the root you eat was once connected to the plant you identified above ground, do not break the plant from its root until after harvesting.

Plants with edible underground parts: air potato, bugleweed, bulrush, burdock, cattail, chufa, curly dock, day lily, duck potato, evening primrose, wild garlic, wild ginger, groundnut, hog peanut, honewort, Indian cucumber root, Jerusalem artichoke, kudzu, meadow beauty, wild onion, Queen Anne's lace, Solomon's seal, sorrel, spring beauty, thistle, toothwort, trout lily.

Plant Greens to Eat

The less we boil or steam edible green plants, the more nutrition we reap from the food. However, bitter tasting greens sometimes need multiple boilings, pouring off old water and adding new to repeat. This bitterness is a signal that the plant contains unwanted chemicals. Sample to determine at what point a cooked plant is palatable. Many plants' green parts are recommended only in early spring before the leaves have accumulated high concentrations of defensive chemicals that can have adverse effects to humans. However, look for new growth at any time of year. New out-of-season growth sometimes occurs because of plant damage or animal predation. In autumn, when the length of daylight mimics spring or when atypically mild weather occurs, there can be "spring growth" into autumn and winter.

Plants with edible green aboveground parts: amaranth, some bamboo, basswood, beech, burdock, cattail, cleavers, corn salad, ox-eye daisy, dandelion, day flower, curly dock, greenbrier, heal all, hemlock tree, honewort, jewelweed, Japanese knotweed, lamb's-quarters, lizard skin, meadow beauty, milkweed, nettle, saw palmetto, pines, broad-leaved plantain, pokeweed, prickly pear cactus, purslane, quickweed, sassafras, smartweed, Solomon's seal, sochani (green-headed coneflower), sorrel, spiderwort, spring beauty, sumac, toothwort, trout lily, twisted stalk, violets, watercress, yucca. **For safety, review proper methods of preparing!**

Edible Flowers

Plants with edible flowers: cattail, dandelion, day lily (buds), elderberry, ox-eye daisy (ray flowers), redbud, rose, spring beauty, some violets, and yucca.

Edible Fruits

Fruits that may be eaten raw or cooked: autumn olive, barberry, blackberry, black cherry, black gum berry, blueberry, chokeberry, currant, dewberry, wild grapes, groundnut, hawthorn, honey locust (pulp inside pod), huckleberry, maypop (in moderation), nannyberry, partridgeberry, pawpaw, persimmon, Chickasaw plum,

black raspberry, red raspberry, redbud, Solomon's Seal berry, serviceberry, sparkleberry, sugarberry, and wild and false strawberry.

Special case fruits with a warning:

Elderberry (blue-black) has enjoyed a good reputation with most foragers, but there have been reports about upset stomachs. Try in moderation.

Hawthorn seeds should be removed from the fruit and discarded. Taste of pulp varies.

Mayapple contains dangerous chemicals throughout. Its fruit must turn from green to a yellow-brown and break off easily from the stalk before harvesting to eat raw.

Young common *milkweed* pods can be dropped into boiling water for a few minutes and eaten.

Sumac berries are used to make the Cherokee drink qualla. Filter before using. (See Chapter 6)

Mulberries must be eaten only when ripe.

Yew berry has edible flesh. Other parts of the plant are very poisonous. Even one seed lodged between teeth can bring on a severe toothache.

The edible red fruits of *prickly pear cactus* are covered with tufts of tiny spines that can work into the skin and become elusive to tweezers or needle.

Nuts and Seeds

Nuts are available late summer through winter – the most common being acorns in fall and early winter. Even acorns that have been sitting on the ground for months (red oak acorns do not root until spring) can be edible. The same is true for some other nuts, so persevere in your search after the prime nut season has passed. A treasure trove of nuts awaits in a rodent's cache. In a survival situation if you find such a supply, steal it with gratitude. The rodent will be better at making up its loss than you are at scraping out a living in the wild, but consider leaving a gift of thanks.

Native nuts are high in saturated fats and without cholesterol. They are also rich in serotonin, known as "brain food" because of its beneficial role as a nerve transmitter during concentration. Nutritionally and medicinally, seeds, including nuts, are the most power-packed part of a plant.

__*Edible seeds:*__ acorns (oaks), amaranth, beech nuts (in moderation), chestnuts, chinquapin, curly dock, grasses, hazelnuts, hickory nuts, hog peanut, jewelweed, lamb's quarters, maple seeds, black medick, pecan, pine seeds, Queen Anne's lace, redbud, wood sorrel, walnuts.

Nut Cracking – Native foragers developed a good method for cracking nut shells. Find a heavy, stable, flattish stone to use as a workbench. On one surface, find (or grind with a harder stone) a divot that allows you to set a nut on end, so that is stands vertically rather than lies on its side. This base is called the "nutting stone." With a smaller "hammer stone" (something comfortable to handle with enough weight that does not require a heavy blow) gently tap the top of the nut by

a blow perpendicular to the nutting stone surface. Start so softly that the shell does not break. Gradually increase the strength of the blow until the shell onomatopoetically says, "Crack!" Learning the exact force to be applied to the strike is most satisfying. The perfect "crack" yields an unbroken nut, or at the very least, larger pieces than does a careless strike.

Pollen

Pollens edible raw, used as a baking flour or soup-thickener: pine, cattail.

"Let thy food be thy medicine and thy medicine be thy food."

~ Hippocrates

CHAPTER 8
The Green Pharmacy
~ Medicines from Plants~

A gentleman living in my county asked me to come out to his land to teach him about the things that he saw around his house every day. Like so many people I have met, he was filled with the earnest and natural desire to understand his home in the broader sense. It is, I believe, one more atavistic need that surfaces from deep in our bones.

In exploring his land and talking about tracks and birds and plants and animal life, he shared with me his long history of producing stones in his urinary tract. On a regular basis, every two months, he felt the formation of a new stone, suffered through it, and then finally passed it by urinating. It was a painful, life-long ordeal that had become routine. His trade-off against the pain and inconvenience was a trophy shelf of jars containing his voided stones.

I told this neighbor that the Cherokees commonly suffered from stone formations. Their oral history is rife with references to medicinal concoctions to rid the body of stones. Their problem was probably genetic-based and due in part to the abundance of oxalic acid in Appalachian plants. When this acid encounters free calcium in the body, it binds with calcium to form a crystal (calcium oxalate) … a stone.

Together we roamed his forests and found a small, evergreen herb, pipsissewa (*Chimaphila maculata*) with dark green lanceolate leaves and a wide white mid-rib vein that gives the plant its striking appearance. (See photo in Chapter 3.)

The herb's common name is a Native American word translated roughly as "it breaks apart the stone." The plant we sought is also called "striped" or "spotted wintergreen," though no such wintergreen aroma can be detected. (There is another plant that bears the name "pipsissewa" – *Chimaphila umbellata*. Both pipsissewas contain a urinary antiseptic, antibacterial compounds, and diuretic properties made available in the form of a leaf tea. Modern science has verified that medicines from both plants reverse stone formation in the body.)

There is a known downside to the overuse, however, and one must partake of the tea with moderation and only when needed. Anxious to try it on the occasion of the formation of his next stone, my neighbor followed an herbalist's guidance on the proper dosage for a tea.

In a few weeks, when he experienced the subtle sensations that marked the beginning of a stone, he drank a modicum of pipsissewa leaf tea for three days. The feeling of the stone's presence disappeared. No pain was experienced. No stone was passed. It had dissolved. He had suffered stones for 30 years, but he produced no more through the next six years that I knew him.

This chapter contains a quick reference to health problems in survival scenarios and the medicinal plants that can be helpful in resolving them. Use the index at the end of the book to read more about these plants. (Some medicinal plants mentioned here may fall outside the scope of survival, but they are included for those who might want to pursue their application with a professional herbalist.) Remember that every person does not react the same way to the same foods and medicines. You will have to learn about your own tolerance to certain chemicals. *Make 100% positive identifications!*

Cross-reference: Maladies or Medicinal Categories and Plants Associated with Remedies

Many books that give the historical uses of plants as medicines overwhelm a reader by showing improbable numbers of remedies for a given plant. A number of published uses are now known to be ineffective or questionable. The following references are those that have been validated by modern research. Refer to Chapter 6 for details of usage. If preparation details for a given plant are not explained in this book, you are meant to pursue its use with a professional.

Adaptogenic – **ginseng** root tea; appropriately mixed root teas of sassafras and yellowroot

Allergy (pollens) – **mullein** leaf mucilage; root rind of **New Jersey tea**; **sassafras** root

Anesthetic (topical) – **cattail** mucilage between leaves

Analgesic – inner bark of **black birch, willow; maple** leaf/bark poultice; leaf of **hawthorn, violet**

Antibiotic – root of **wild garlic, Queen Anne's lace** (see warning in Chapter 6); leaf of **agrimony, heal all, mullein**

Anti-inflammatory – leaf of **agrimony, jewelweed, mullein, broad-leaved plantain, yarrow**; root of **ginger, yellowroot**; fruit of **wild grape**; inner bark of **willow**

Antiseptic – inner bark of **beech, hemlock**; leaf of **broad-leaved plantain**; sap of **oak, pine, sweetgum; juniper** berry

Astringent – leaf of **agrimony, bugleweed, chestnut, cinquefoil, ragweed, witch hazel**; inner bark of **hemlock, beech, oak, redbud**; root of **New Jersey tea, wild geranium, yellowroot**

Bleeding – root of **alumroot, wild geranium**; leaf of **yarrow**; mucilage between **cattail** leaves

Bronchitis – leaf of **mullein**; inner bark of **white pine**; sap of **sweetgum**

Bruising – select **violet** leaves

Burns – leaf of **cleavers, maple, strawberry**; root of **greenbrier, Queen Anne's lace, yellowroot**; inner bark of **maple**

Cholesterol level – root of **garlic, ginger**; fruit of **hawthorn**; seeds of **broad-leaved plantain**

Cold (prevention) – root of **garlic, sassafras**

Cold (treatment) – root of **garlic, New Jersey tea**; leaf of **hemlock tree, pine, yarrow**

Constipation – leaf of **ash** tree, **mints, wild lettuce**

Diarrhea – leaf of **alumroot, blackberry, strawberry**

Diuretic – leaf of **black birch, dandelion, heal all, Joe Pye weed, pipsissewa**; root of **Queen Anne's lace**

Ear infection – flower of **mullein**

Energy loss (mental and physical) – **ginseng**

Expectorant – leaf of **mullein, yarrow**; inner bark of **white pine**

Gas – root of **ginger**; seed of **Queen Anne's lace**

Fever – inner bark of **flowering dogwood**; leaf of **mint, yarrow**

Food/water poisoning – root of **sassafras**

Headache – inner bark of **black birch, flowering dogwood, willow**; leaf of **violet**

Hemorrhoids – leaf poultice of **broad-leaved plantain, witch hazel**

Histaminic swelling – juice of **jewelweed**; root rind of **New Jersey tea**; **broad-leaved plantain**

Hypertension – root rind of **New Jersey tea**

Hypoglycemia – root of **Solomon's seal, Jerusalem artichoke**

Indigestion – seed of **broad-leaved plantain**; leaf of **mint**; root of **ginger, yellowroot**

Infection (wound) – leaf of **broad-leaved plantain**; root of **curly dock**; leaf of **maple, mullein**; sap of **pine, sweetgum**

Insect bites and stings – leaf poultice of **curly dock, jewelweed, broad-leaved plantain, pale Indian plantain, bear paw rosinweed** (*Silphium compositum*)

Nausea – root of **ginger, sassafras, yellowroot**; leaf of **wood sorrel**

Poison Ivy, Oak, Sumac rash – root of **alumroot, New Jersey tea**; bark of **beech, black birch, hemlock, oak**; leaf of **jewelweed, broad-leaved plantain**; and **poison ivy** (see Chapter 10)

Poultice – (topical, to draw out foreign matter) – heated sap of **pine**; leaf of **pale Indian plantain**

Sedative – best to consult an herbalist for these: bark of **black birch; flower of red clover, hawthorn, mullein**; root of **black cohosh**; leaf of **holly**; fruit of **maypop**

Skin fungus, rash – root of **bloodroot, New Jersey tea**; leaf of **curly dock, jewelweed, mulberry;** topical twig tea of **spicebush**

Splinter – (see poultice)

Stinging Nettle rash – leaf of **curly dock, jewelweed, broad-leaved plantain**

Stomach upset – leaf of **mint, sorrel**; flower of **mullein** (for a virus); root of **yellowroot**

Stress – root of **ginseng** or **yellowroot/sassafrass** by an herbalist

Styptic – root of **alumroot, curly dock, New Jersey tea, geranium**; leaf of **yarrow; cattail** gel

Sunburn – topically applied leaf tea of **cleavers, curly dock,** or **wood strawberry**

Throat (sore) – gargles: leaf tea of **alumroot, agrimony,** bark tea of **beech, hemlock, New Jersey tea, oak, redbud, sweetgum**; root of **yellowroot**

Vomiting – greenery of **wood sorrel**

"You boys come draggin' back in here ever' night to chow down and replace whatever got broke or wore out. Hell, them Injuns out there live off'a ever'thin' you just turned your back on."

~ *Short Stuff, cook for New Surrey Downs Ranch*, Indigo Heaven

CHAPTER 9
A Lifetime Supply of Everything You Need
~ *the Utilitarian Lore of Plants* ~

Learning about the many gifts in Nature is more than a pragmatic reservoir of forest lore. It is learning to integrate with your natural surroundings on an intimate level. It is about getting comfortable … experiencing contentment on the most basic level. It is about having fun in the original playground of the world and, while there, learning the history of the ages. It is about getting our children into a healthier place … and, once there, being awestruck.

The paleo-child is still hidden deep inside the genes of our young ones (just as it is, hopefully, inside every adult). That child of Nature can still be tapped and released, and along the way if that child learns of and uses the ancient gifts that still abound in the wild – the same utilitarian gifts known to the Cherokees of yesteryear – then he or she will grow closer to it.

What follows is a reference list for categories of Nature's gifts that are most often needed in a survival situation. Refer to Chapter 6 for plant taxonomical names and descriptions.

UTILITARIAN PLANTS
Cordage
(See Chapter 13 for instructions on making rope)

<u>*From inner bark*</u>: ash, basswood, cedar, cherry, cypress, grapevine, hawthorn, hickory, juniper, kudzu, black locust, mimosa, mulberry, Osage orange, pawpaw, tulip magnolia, walnut, willow

From roots: elm, fir, hemlock, hickory, juniper, locust, mulberry, pine, spruce, walnut, willow

From leaf fiber: cattail, grasses, yucca

From stem: blackberry, dogbane, grapevine, kudzu, milkweed, ironweed, nettle, velvet leaf

From animals: (covered in *Secrets of The Forest, Volume 3*): **mammal skin, sinew, hair,** and **gut**

Dyes (a sampling of easy-to-find sources)

alder, dandelion, elderberry, hemlock tree, mullein, pokeweed, walnut, yellowroot

Insect repellents

Expect varying results with different individuals – depending on body chemistry, diet, and fitness. (Refer to Chapter 6 for instructions on tolerance-testing and usage.)

General – Campfire smoke, mud

Leaf – ash, beautyberry, bracken, hazelnut, hickory, juniper, pawpaw, perilla, sassafras, walnut

Root – bloodroot, garlic

Flowers – black cohosh, elderberry, ox-eye daisy

Seeds – broad-leaved plantain, pawpaw

Wood – green chips of juniper or sassafras

Natural Soaps

A soap covers whatever it touches with a film that improves the rinsing process by like-charged particles repelling one another. Saponin, a natural soap, can be found in these plants:

Leaf – yucca, soapwort, jewelweed, sweet pepperbush

Root – pokeweed, New Jersey tea, yucca, jewelweed, soapwort, sweet pepperbush

Stem - jewelweed, soapwort

Woods for Bow-Making

White oak, dogwood, hophornbeam, juniper, cedar, maple, Osage orange, white ash, hickory, black locust, white mulberry, sassafras, birch, yew,

yaupon holly, bamboo, elm (Bow and arrow making is covered in *Secrets of The Forest, Volume 4.*)

Arrow Woods – high bush blueberry, box elder maple, swamp dogwood (*Cornus oblique*), **arrowwood** (*Viburnum dentatum*), sourwood, rivercane, cypress, black locust, Osage orange, reed, chokecherry, serviceberry, rose, witch hazel, chestnut, plum, fir

**Arrow/Dart Poison** (Cherokee) – **bulbous buttercup** (*Ranunculus bulbosus*), yucca

**Fish Poison** – perilla (whole plant); seeds of **buckeye, mullein;** husk of **walnut**; root of **yucca**

**Glue** – Hide scrapings heated to a glob; melted **pine** pitch mixed with hardwood ashes

**Fire-Making Trees and Weeds** – (Fire by friction is covered in *Secrets of The Forest*, Volume 2.)

Bow-drill: ash, aspen, bamboo, basswood, box elder, birch, buckeye, catalpa, cedar, china-berry, cottonwood, cucumber tree, cypress, elm, fir, hackberry, hemlock, hickory, juniper, maple, mimosa, pawpaw, poplar, redbud, redwood, sagebrush, sassafras, short-leaf pine, sumac, sweetgum, sycamore, tamarack, tree of heaven (ailanthus), tulip magnolia, willow, white pine, cracked-cap polypore bracket fungus

Hand drill: cattail, elderberry, evening primrose, goldenrod, horseweed, lamb's quarters, mullein, trumpet creeper vine, yucca, dead exposed roots of **hemlock tree** and soft hardwoods … and virtually any dry, woody weed's winter remnant, if it is strong enough to hold up to the rigors of drilling on a hearth.

Tinder sources

**Dead inner bark:** ash, basswood, cedar, cherry, cypress, juniper, black locust, oak (rarely), poplar, tulip magnolia, walnut, willow

**Dead leaf fibers:** cattail, mullein, pine needles, **grass**, and many **deciduous leaves** that have decomposed to their net of veins

**Pappus:** aster, cattail, cottonwood, dandelion, wild lettuce, milkweed, thistle

"The poison oak or poison ivy so abundant in the damp eastern forests, is feared as much by Indians as by whites. When obliged to approach it or work in its vicinity, the Cherokee strives to conciliate it by addressing it as 'My friend' (hi' gǐnaliǐ)."

~ James Mooney, Myths of the Cherokee and Sacred Formulas of the Cherokee

CHAPTER 10
Poison Ivy

During a plant class with twelve-year-olds, I was kneeling before a specimen of poison ivy (*Toxicodendron radicans*) and explaining what the students needed to know about it: how to recognize it, what to do if they knew they had been exposed, and how to cure the rash. When I was finished, one student announced proudly, "I'm not allergic to it. I can touch it all I want."

As I was in the middle of explaining that such immunity can change at any time in a person's life, the student, smiling broadly, plucked a leaflet and began rubbing her arms with it. We all watched with mixed reactions of doubt, horror, awe, and curiosity.

The next morning she came to me as the very portrait of misery. Her arms were covered in blisters. "I itched all night," she whimpered. We walked together to the river to harvest jewelweed, and her problem began to resolve immediately. Her pride took a little longer.

Most of us are born into this world immune to poison ivy. But that usually changes, even if it's as late as our adulthood. When that happens, it almost always means that the allergy is permanent. However, there are contrary claims: Some people, who as children suffered a severe rash, profess to have developed immunity as adults. And others, like those of African descent, might never become allergic. Contrary to old-husband's tales I have heard, Native Americans were vulnerable to poison ivy.

There are many other plants that can cause dermatitis to humans: **black-eyed Susan, dandelion, daisy fleabane, bloodroot, virgin's bower, buttercup,**

pokeweed, Queen Anne's lace, horseweed, mullein, yellow pine, pawpaw, and **Virginia creeper,** to name but a few. These just-named plants are not universally problematic like poison ivy and its kin. I have handled all of the above, and, in my particular case, only the ones with "poison" in their names affect me. Of course, sensitivity to plant irritants varies from person to person.

Poison oak and poison ivy look so similar that the layperson should consider them one and the same.

Contracting poison ivy rash requires coming into physical contact with the plant or its offending chemical – an oil named "urushiol," which is manufactured in the leaves and then distributed to all other parts of the plant. Contact is most often made by touching the leaves, fruit, root, or vine but can occur from water dripping off a vine or by exposure to smoke-borne resin.

The rash usually spreads the same way, physically, by transferring oil from one locale on the body to another by way of scratching fingers, clothing, etc. If your dog has been sleeping in poison ivy just before you give her a hug, prepare for a mysterious outbreak along your arms and cheek. But there is more to the story than topical transference. In some cases the malady is known to have become systemic, pervading the body internally.

Sometimes we suspect poison ivy rash, when actually the problem is tiny critters like chiggers, mosquitoes, ticks, and fleas, which covertly insert mouthparts through the skin and induce itching. Since we often don't see the intruder, we might be inclined to accuse a plant.

Virginia creeper is a vine that is poisonous to eat but not necessarily dangerous to touch. In the fall, when the leaves have turned to stunning scarlet and gold, this vine – also called "woodbine" – takes on irritant qualities to certain people who are sensitive to it. It is this plant, more than any other, that I have heard people incorrectly assign the name "poison oak." Poison oak and poison ivy are so similar that the layperson should consider them one and the same. And unlike poison ivy or poison oak, Virginia creeper has five (not three) leaflets, which are palmate (not pinnate). It produces blue berries, not whitish/gray like poison ivy.

Virginia creeper

Another plant often mistaken for poison ivy is a very young box elder maple. Its telltale identifier is its opposite arrangement of leaves. (See the color photo section under "maple.")

Anatomy of a Rogue Plant

Poison ivy is a compound plant, each of its three leaflets being ovate. Therefore, the old adage "leaves of three, let it be" should read: "leaves of three leaflets… (To avoid confusion, the term trifoliate is sometimes used.) But if you follow either folksy warning, you're going to miss out on lots of other plants that typically show three leaflets, like black medick, hog peanut, and young blackberry canes, all of which have edible parts.

Are trifoliate compound leaves pinnate or palmate? Three leaflets may not suggest a feather-shape as, say, the longer rows of leaflets on a locust or a sumac, but if the three leaflets are all trying to point more or less in the same general direction (toward the tip of the compound leaf), then the leaf is pinnate … like poison ivy.

Palmate leaves (like Virginia creeper's) spread their leaflets in multiple directions from a common center, distributing them more equally among the 360° available.

No one can render one drawing or take one photograph or point out one specimen of poison ivy as a reliable guide to identification. The plant

poison ivy

has too many guises. Namely, this has to do with its teeth, which can be numerous, sparing, or altogether missing. Thankfully, there are a few other nuances of the plant's physiology that are useful in identification.

Where the leaflets' edges neighbor one another, their margins tend to be smooth. Serrations occur usually on the two lateral leaflets on their outermost margins. If teeth develop on the middle (terminal) leaflet, they show up close to the tip. Usually.

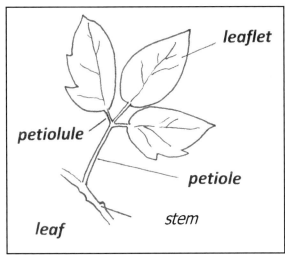

poison ivy

Each lateral-leaflet is asymmetrical, with more leaflet-blade material showing on the outside of its midrib vein than the inside. The terminal leaflet is symmetrical.

The leaflet stalk (petiolule) of the center leaflet appears longer than the other two. A burgundy-red pigment usually marks the point of the three petiolules' convergence. In autumn when the compound leaf drops away, the remaining scar is slightly raised from the stem like a tilted shelf. (Most leaf scars on other plants lie flat against the stem, like a brand seared into the bark.)

Poison ivy's small, green, five-petal flowers emerge from the leaf axil in late spring and early summer. After flowers are pollinated they fruit as gray-white berries. Mature vines climb trees and boulders and develop a thick covering of air roots but no tendrils. Other vines have a "hairy" appearance too, however, like climbing

> *No one can render one drawing or take one photograph or point out one specimen of poison ivy as a reliable guide to identification.*

hydrangea. But hydrangea leaves and branches grow opposite one another. Poison ivy leaves (not leaflets) and branches grow in an alternate fashion, never opposite. Even in winter you can look at the bare branches of a vine and distinguish this by the leaf scars. You'll learn about these botanical markings in Chapter 11.

When poison ivy leaves first emerge in April they may appear pale green, red, silver, or any combination of these colors. As they mature the green color deepens and dominates, and the surfaces of leaves take on a varnished sheen. If a leaflet is snapped off, a milky sap beads up at the break. As poison ivy grows it can assume the shape of an herb, a shrub, or a vine. When a vine – because of its far-reaching branches extending sometimes eight or ten feet from a tree trunk – the branching might be mistaken for limbs of the supporting tree trunk.

How Does Poison Ivy Make Us Itch?

When poison ivy's oil (urushiol) contacts us, our immune system triggers a chain of events that usually leads to bothersome itching at the least. But extreme cases are quite serious. When the problem goes internal, the reaction can result in death. Topically, it takes about fifteen minutes for the oil's molecules to bind with protein in the epidermal cells. If the skin area is callused, say soles of feet and palms, it takes longer or might fail to bond. Because urushiol is water-soluble it can be quickly washed off, but once the oil bonds to your skin, it's too late.

What does urushiol actually do? In small dosages, like touching a plant, it may do very little. It's how our bodies respond that brings on the misery. The immune system over-reacts. Why? Because in large quantities urushiol carries the potential to suppress the body's production of prostaglandin, a fatty acid that aids in the body's battle against inflammation. The human immune system goes on high alert when confronting an oil with this kind of potency. Even with minimal exposure to the oil, a fully-blown immune reaction can follow. The severity of the reaction varies from person to person. The immune "cavalry" of white blood cells and lymph crowds into the affected cells to do battle, causing blistering and tightening ... and itching.

The itching of urushiol-bonded skin can occur within two hours or less after contact. Or it could appear a day or more later. It all depends upon a person's particular sensitivity.

Viable urushiol can survive on the surface of objects indefinitely – depending upon its exposure to rain. Century-old, dead, and sheltered vines have been found saturated with urushiol.

The rash that occurs from contacting poison ivy can, with some people, develop into a systemic problem – spreading internally through the body. When that happens the situation becomes very serious, affecting T-cells in the immune system. I have never experienced this side of the malady, but I have seen sobering pictures of victims covered with sores. Such an outbreak requires professional help.

More commonly, the rash is spread inadvertently by topical transfer of urushiol to a new locale on the body. This might occur by contact with contaminated clothing or with pets or simply by hand after scratching an itch. The liquid that

oozes from the blisters of a rash is water-based and can dissolve any free urushiol at the site of the rash. Such "urushiol soup" can then spread and expand the afflicted area as well as be transferred to another site. That's why it's important to wash the rash even after the blisters appear.

How do we protect ourselves?

1. **_Avoid contact_** with it. The easiest poison ivy preventative is being observant enough to step around it.

2. After contacting it, **_wash off immediately_**. Urushiol is water-soluble, so it's an easy task to remove it. Soap is not necessary, though soap makes the job easier because it binds with the oil, making it easier to rinse. But how you rinse is important. If you cup water in your hand and spread that water on the exposed area, you'll be making "urushiol soup" and worsening the problem by expanding the range of contact. Instead, submerge the exposed body-part in a creek and scrub, letting the current carry away the oil. If you wait too long to do this – say, a half-hour – no amount of washing will rid you of the oil that has already bonded with the skin. But it will wash away un-bonded oil, which might accidentally be transferred. If you use soap, spare the creek any soap pollution by first rinsing away from the stream before a final rinse at the creek.

3. **_Never walk through rain-soaked poison ivy._** Rain bruises the leaves and releases the oil. Water standing on the leaves dissolves urushiol and can pass through porous clothing to contact the skin. If the clothing is not porous – like rain gear – you'll be carrying urushiol, which can drip down into your socks. If ever you use a tree for shelter during a rain, always examine the tree to check for climbing poison ivy vines. Make sure you are not standing in a urushiol shower.

4. Apply a commercial **_urushiol shield_** to the skin. These can be found at sporting goods stores. I can't speak for their efficacy, because I've never used them.

5. **_Never use an axe or chainsaw on a poison ivy vine_**, as oil can spray into the air. If you insist on such a project, dress in protective clothing and safety glasses and plan to wash immediately.

6. **_Never burn poison ivy!_** Smoke carries the oil, which can then find its way to your skin by airborne means. Or worse, it can be inhaled to inflame the lungs, a condition that can prove irreversible and fatal.

What to do when you get the rash

When the rash appears, wash/rinse the area thoroughly. Although oil that has already penetrated the skin cannot be washed away, any excess oil that has not yet bonded with the skin needs to be removed so as not to be transferred. Consider all potential carriers: clothing, fingernails, tools, tent, dog, etc. Once you have washed everything, it's time to apply medication.

Many Native American poison ivy medicines were prepared as bark teas and applied topically to the rash – black birch, for example, whose bark contains wintergreen oil (methyl-salicylate), an analgesic. Tannin (or tannic acid) is another

chemical that can be applied. It's a common astringent of Eastern forests that can be found in the root of New Jersey tea (a shrub), in the bark of oak, beech, and hemlock, in acorns, and in oak and chestnut leaves. The crushed greenery of agrimony, cinquefoil, sanicle, and ragweed are other sources. Tannin draws up tissue, squeezing the immune fluid out of the blisters. It is also antiseptic, which provides a bonus by preventing infection of overly-scratched blisters.

> *Inhaling poison ivy fumes from a brush fire often leads to death.*

Even considering all that is available from drug stores – calamine lotion, Benadryl, hydrocortisone cream, diphenhydramine, astringents in alcohol, etc. – nothing works better than jewelweed, a wild impatiens, also called "touch me not" for its "exploding" seedpods. This succulent plant grows in wet, sunny areas, especially along the banks of ponds, large creeks, and rivers. Its mucilaginous fluid can easily be crushed from any part of the plant, but especially from the stem (before the plant toughens in late summer.) This juice contains an anti-inflammatory agent, alleged to be kinder to the adrenals and immune system than cortisone synthesized in a lab. Smear the mucilage into the rash, using the crushed plant itself as an applicator. Relief is immediate. Apply three times a day, each time using fresh material. (Jewelweed cannot be stored except by freezing the plant blended in water as a medicinal ice cube.) Within a day and half the rash is noticeably subsiding and, throughout the treatment, itching abates. By day-three you could be itch-free.

Broad-leaved plantain is also very effective in healing the rash. Use just as you would for stings.

Making yourself immune

Though there may be a risk involved for certain people, many outdoor enthusiasts create an annual immunity to poison ivy by eating one young leaflet once a week for seven to nine weeks. I have been using this method for many decades and have never experienced a problem.

Each April, when the leaves emerge from their buds, I ingest one tender leaflet. At this stage it is usually reddish and/or silver – the size of my little fingernail. I repeat this on a weekly basis – one leaflet per week for nine weeks. As spring progresses I seek out blades only slightly larger each time. By the end of the process my ninth leaflet is solid green and as large as two thumbnails.

The first time I deliberately broke off a leaflet to eat and eyed the milky sap that beaded up on the broken leaflet stalk, I felt I was taking a leap of faith to insert it into my mouth. I knew of three people who had used the immunity method successfully and without ill effects, and because I trusted them I took the plunge and hoped that I would not be an exception to the rule. I placed the leaflet on my tongue, chewed, swallowed, and right away began to imagine an invasive rash on my tongue, esophagus, stomach … and who knew where else down the alimentary canal! But all went well. Nine weeks and nine leaflets later, I ran a test. I picked a mature poison ivy leaflet and rubbed it into the inside of my forearm. I did not wash it off. After an hour a single flesh-colored bump (not a typical pale blister) rose up on my skin. Not once did it itch. Fifteen minutes later, it disappeared. No other signs appeared.

A warning: I, of course, cannot guarantee the safety of this method for any one person. I am simply sharing my experience. Through the grapevine of wilderness skills teachers I have learned of two persons who developed adverse effects after eating that first leaflet; therefore, it must be said that there is a gamble involved. Though it seems to be rare, the possibility for a backfire does loom. I have repeated this homeopathic immunization every spring throughout my adult life, as have dozens of my friends. None of us has ever experienced a negative reaction.

This annual immunity lasts for me from late spring through midwinter. Only once did the method fail me. That was the year that I developed a rash in early spring before my nine-week installments were complete. Though I finished out the doses, I was not immune that year.

Does this plant have value?

To be frank, to ask such a question is a bit narrow-minded. It alludes to a presumption that all things in Nature are designed for the benefit of humans. Ecology is an extremely complicated system of interactions. Though it may seem that we humans take center stage in the affairs of the world, remember that the Earth fared very well through the eons before the arrival of mankind.

Plants continue to be crucial players in the life-on-Earth equation. As a green plant, poison ivy is one more green ally that captures the sun's energy for use on our planet. Two of the by-products of this miracle that we call "photosynthesis" are the oxygen that we breathe and the recycling of the carbon dioxide that we exhale. I've watched deer and other animals eat the leaves of poison ivy. (This could be fatal to a human.) And I've seen raccoons, songbirds, and crows eat the berries. The flowers produce a nectar favored by honeybees. If you could ask those creatures about the worth of poison ivy you might get a more balanced answer.

Plants have evolved interesting methods by which to protect themselves from predation – like thorns and highly toxic chemicals – and we have admired those defenses to the point of emulating these plant features in our inventions of barbed wire and pesticides. But we seem to take poison ivy's defense system personally and feel persecuted to acquire its wrath when we take an innocent walk in the woods. Think of it this way: poison ivy is a teacher. Like venomous snakes, this unique plant sharpens our observation skills.

"*If you know how to listen, the Standing People will talk to you. In Spring they wake up and begin to whisper. In Summer the words rush with the wind through the leaves. Autumn brings the falling asleep song, until in Winter there remains only a soft breathing.*"

~ *Black Otter Woman*, Legend of the Medicine Bow

CHAPTER 11
Winter Botany

Recognizing trees in winter might seem an improbable proposition to some, but there is quite a lot of identifying evidence to seek out in the dormant season. Bark patterns, limb structure, growing site, and remnants of leaf blades and fruits are all clues that can be enough for an experienced eye. But there are smaller clues that are even more exacting for the identification process. Discovering them can open up a new world of Nature study, one which for all your life has brushed your jeans or was hidden under your grasp as you pushed aside a winter branch.

This chapter will deal with very small things – tree and shrub secrets that can be thought of as botanical "fingerprints." We will also consider larger (though inconspicuous) bits of evidence, such as brittle, withered remnants of dead herbs. Most of the general public is not even aware of the details that you are about to study. Once you have become familiar with the anatomy of winter botany through this chapter, to make use of it you will need a winter key – a book unto itself. My regional books of choice are: *Identification of Southeastern Trees in Winter* by Preston and Wright and *Trees of Georgia and Adjacent States* by Brown and Kirkman. It would be helpful to carry a magnifier for this exploration of minute details.

First on the agenda is a broader concept for which we must take a step back to see a grander scheme of botany: spirals. In plant growth there is an underlying corkscrew theme. Consider a tree branch with alternate leaves. Following that branch outward you will discover that its leaves emerge in a spiraling pattern. If you tied a string to the first leaf stalk and stretched the string to the next leaf stalk and the next and so on … that string would twine around the branch in corkscrew fashion. This spiraling of leaves allows for maximum absorption of sunlight.

Each tree genus follows its own specific pattern. For example, all beech trees make that leaf-to-leaf spiral in 120° intervals. If you viewed a branch from its end, three successive leaves would equally trisect the 360° available in three *ranks*. The next leaf (fourth) would fall in line with (on the same rank as) the first as the trisection starts anew.

A plant's spiraling formula is true not only for leaves but also for buds, branches, twigs, cone scales, bud scales, and disk flowers at the center of daisies and sunflowers. However, this orderly system is often distorted by natural factors – like a twisting grain in the trunk, abnormal growing conditions, animal predation, or natural accidents.

If a tree's alternate leaves spiral in 180° intervals, the branch takes on a planar look, the leaf blades on both sides creating a flat bough. Basswood, elm, pawpaw, and birch share this trait and so are *two-ranked*.

Most plants with opposite leaves rotate their successive pairs (from one node to the next) by 90°, making them four-ranked. Examples are maples, buckeyes and most mints.

Look down the stem of a whorled herb and you will see that successive whorls rotate enough to allow a lower whorl to catch the light that slips through the whorl above it. With this principle of the spiral in mind, let us forge ahead into the intricacies of winter plant identification. To get a firm grasp of the big picture, we will first discuss a plant's behavior in spring.

During this time of emergence, our eyes are drawn to the opening buds of leaves and flowers, but another botanical effort is underway. As a bud opens, just above it in what will become the *leaf axil* (the upper angle between the leaf stalk and the stem or branch from which it emerges) look for another bud that the plant has begun to manufacture for the *following spring* – a bud that will open one year later. In this nascent period, the future bud is miniscule. But in mid- to late summer when the plant has completed its growth for the present season (twigs, leaves, flowers, fruit), its energy is directed toward those tiny buds, plumping them up to full size at which point they will lie dormant through fall and winter until the next spring.

All buds do not open. Some are held in reserve in case opening buds are lost to disease, animal consumption, accident, or a late freeze. (A researcher once counted all the buds on a mature elm tree … approximately seven million! The elm would only make use of a fraction of these.) All buds, whether they open or not, are complete units neatly packed with just the right amounts of gases, water, insulation, packaging, and antiseptic for freshness.

Unlike so many animal embryos, a baby leaf (encased in its bud) looks very similar to its future adult self, only in miniature. It is challenging to undertake the delicate probe into a bud, prematurely opening a furled or folded leaf, but with skillful fingers it is sometimes possible to identify a tree by exposing its embryonic leaf.

You will find that trees have various methods of packaging new leaves inside their buds. A tulip tree leaf is folded in half along the mid-rib vein. Birch and

beech leaves are folded along the pleats of their ruler-straight lateral veins. Other leaves are twisted, crumpled, rolled, or flattened against other bud parts.

Journey into a bud
— In winter or early spring locate a tulip magnolia tree (*Liriodendron tulipifera*) that has not yet opened its buds. Select one of the larger terminal buds (capping the end of a branch) and carefully peel away the two halves of its outer green covering. At first you may believe you have unwrapped only another bud. Look more closely. A tiny embryonic leaf is pressed into the outside curve of this inner bud. (If this outermost leaf succumbed to winter, it may appear as a tiny maroon corpse pressed to the newly revealed bud that lay underneath the first bud. If so, open this next bud to expose a healthy nascent leaf.) The leaf is folded in half lengthwise and connected to the embryonic twig by a most slender and fragile petiole. Gently pry the leaf away from the bud without breaking the petiole. Then with surgical care and patience, wet the leaf with saliva, massage it gently between thumb and forefinger, and attempt to unfold the leaf. (It can be done!) You are now ready to open the next newly revealed bud and repeat the unveiling of another new leaf and yet another bud. These multiple leaves emerge that spring along the new branch material that is also encased inside the larger terminal buds.

Buds

Buds come in different shapes and sizes. That's one factor that helps in the identifying process. Because *terminal buds* are usually larger than *lateral buds*, they make better specimens to study. Some trees, like flowering dogwood, possess two entirely different shapes of buds – a plump one (resembling an onion) for flowers and a thin one (like a bird's beak) for leaves.

The covering of a bud offers another clue as to the genus of a tree. Bud-envelopes (*scales*) can be comprised of two sealed halves like an elongated clam. Such a scale pattern is called "*valvate*." You peeled away valvate bud scales in the last hands-on exercise. Or buds can be covered by roof-like shingles of scales, an arrangement called "*imbricate*." Note that opposite-leafed plants have opposite imbricate scales while most alternate-leafed plants have imbricate scales that follow the same spiral formula as their leaves. Some buds are capped by a single conical scale that resembles Merlin's hat. Willow and sycamore are examples. If you search for sycamore's "next-spring" buds before the present season's leaves have fallen away, you may be mystified. Sycamore buds are hidden *under* the base of a leaf's petiole. When that leaf finally sails away in fall, the new bud will be visible – which explains why, when you pick up a large handsome sycamore leaf off the ground, its petiole has a swollen end with a conical concavity ("Merlin's hatbox"). Some genera of trees show naked buds with no covering of scales at all, like pawpaw and witch hazel.

Since the terminal bud caps off the end of the branch, the ringed scar that it leaves can mark that year's growth limit. By looking along a branch and counting these rings, you can read a branch's age, gauging those years in which growing conditions were either good or poor, according to length of growth. But don't confuse this ring with those trees whose lateral buds' bases completely surround the twig and leave ring-scars at every node.

Such buds are called *"amplexicaul."* Sycamores and magnolias (like the tulip tree) demonstrate such multi-ringed branches. The piece of bud that creates this scar is called a *"stipule,"* which is a part of the bud's protective envelope that falls off from the base of the petiole after the new leaf has emerged.

Stipule scars on many trees are missing altogether (persimmon, hickory, sassafras) or quite small (basswood, mulberry). Sometimes they are visible on tree <u>trunks</u> where limbs once emerged (tulip tree, birch, maple), but instead of encircling the stem they form inverted V's. (See the tulip tree photo in the color photo section.)

amplexicaul bud rings on a tulip tree twig

If you find 1"- or 2"-long bumpy twigs, each with many terminal bud scars compressed into its short length, you have found *spurs*, dwarf twigs which allow a tree to fill in its inner orb with leaves. A two-inch spur might be a dozen years old. Count the rings and note the compactness of spurs on tulip trees, sweetgums, and birches.

Another confusing variable when accounting for yearly growth on a branch is the fact that some trees have no terminal bud – like sycamore, birch, sourwood,

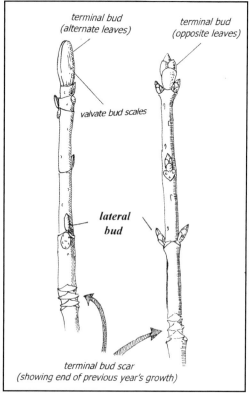

terminal buds and leaf scars

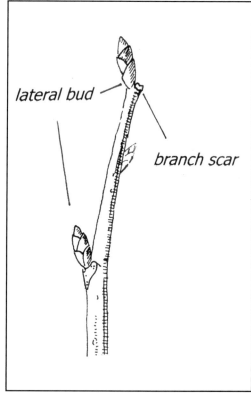

lateral bud/branch scar

silverbell, persimmon, and sumac. Should you go in search of one of these trees to verify this lack of terminal bud, you might feel confused, for you'll see a bud at the end of every branch.

Look closer and you will trade confusion for academic satisfaction and the beauty of intricate inspection. That so-called "terminal bud" is really a *lateral bud*, the last one the branch made in its growing season. In time the branch died back to that bud, broke away and left a *branch-breakage scar*. That scar is smaller than one might expect, because the bark tends to tighten around the wound. This non-terminal bud at the end of a twig will probably demonstrate an angle from the twig rather than capping the twig like a linear extension of the twig itself. Usually, branches without terminal buds exhibit a zigzag growing pattern between nodes.

Leaf Scars

In the dead of winter you can expect to find an array of intricate markings where a leaf once attached to a twig or branch. These botanical "fingerprints" form because of the intimate connection shared between a former leaf and its branch (due to water/sap flow). In the autumn a layer of corky tissue seals off the petiole from the branch, blocking the influx of water and nutritional needs. This event is called "*abscission*." The leaf dries, becomes brittle, and eventually snaps away at this dividing layer, which now remains as a scar on the wood. If you could find the very leaf that fell from that node to compare the base of its petiole to the scar, you would see matching mates somewhat like an electrical outlet and an appliance plug that fits into it.

The outline of the *leaf scar* on the wood is a major consideration in plant identification. Here are a few possibilities for that shape: heart, circle, oval, horseshoe, crescent, shield, triangle, V, and half-round. These leaf scars can be used to determine ranks and spiral formulas just as if they were leaves. Inside these leaf scars are more tree "fingerprints" that can be used for identification.

The tree's sap conduit fibers (from which rope can be made in some species) are visually hidden to us in living trees until they emerge as veins in a leaf. The interface of petiole and branch, where these fibers enter the leaf, reveals a pattern of bundle traces that can be seen on both petiole and leaf scar, one showing tiny bumps, the other tiny concavities. Discerning the shape and pattern of *bundle traces* is more effective with a magnifying lens.

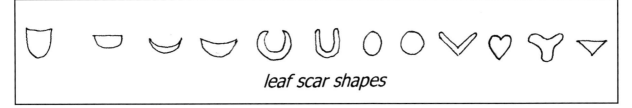

leaf scar shapes

More Buds and Scars

Above the leaf scar (usually) sits the new bud that will take the fallen leaf's place. (Sometimes there are additional accessory buds clustered around this primary one.) Buds open up in varying styles. Some bud scales dissolve with the flow of spring

sap. Others come unglued at the seams. Some simply loosen and fall. A few separate and rotate on their bases like crank-operated windows or shutter slats. Others rupture.

Here are a few notes on exceptional buds that might help as you begin your study: Oaks tend to cluster many buds at the tip ends of their quite straight outer twigs. Dogwoods have two types of buds: bird's beak and onion-shaped. Willow has a conical, one-piece, seamless bud. Sycamore buds hide beneath the petiole. Locust buds are often scabbed over by bark and difficult to find. Beech buds are long coppery spears. Blueberry buds are like opened birds' beaks. Spicebush buds often appear in pairs or quartets.

If present, *stipule scars* can be inconspicuous slits or folds in the bark extending from the top of the leaf scar like semi-drooping wings. Or they could be obvious, encircling the twig. Stipule scars are problematic to the novice. The same tree may exhibit stipule scars at some leaf scars but not at others. That is why it is always prudent to look at several specimens of scars when studying any given plant. If, out of five examinations of five different buds, only one stipule scar is found, categorize that tree as having stipule scars.

All along the bark of a twig, branch or trunk, look for *lenticels* – tiny or noticeable pores through which gases are exchanged. Lenticels may appear as dots, holes, slits, raised dashes, or

> **It is always prudent to look at several specimens of scars whenever studying any given plant.**

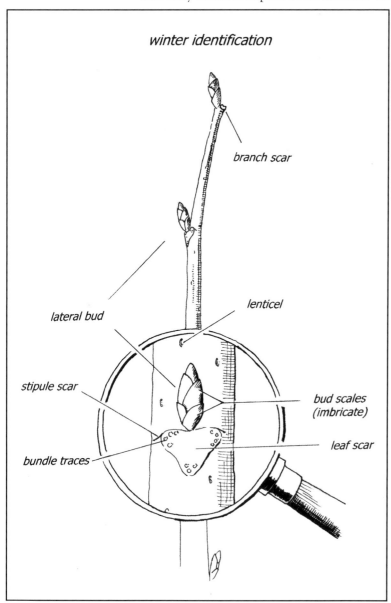

winter identification

branch scar

lateral bud

lenticel

stipule scar

bundle traces

bud scales (imbricate)

leaf scar

birch lenticels

tiny discolorations. Leaves, bark and even roots must breathe – which is why some trees can drown in shallow water. Sometimes to compensate for too much ground water, a tree might swell the base of its trunk above the waterline to increase its surface area and, hence, its number of lenticels. Tupelo and hickory do this. Cypress roots exhibit a clever solution to prevent drowning by dispatching "knees" upward to the surface as lifelines for air. Many maples growing in swampy ground are covered with giant, warty protuberances, another air-compensatory addition of surface area. (These "warts" make handy climbing rungs.)

<h2 style="text-align:center">Pith</h2>

Inside a twig, branch, or trunk lies a telltale *pith* that runs through the core of the wood. This pliable and porous core allows for water storage and for trunk flexibility in high winds. (Herein lies the philosophical metaphor about those who can bend to pressure – rather than break – being the stronger.) A very sharp knife can cut a cross-section in a twig to reveal the shape of the pith's cross-section – another

identifying characteristic of trees and shrubs. (A dull knife will crush the pith and alter its true shape.) Pith cross-sections can be circular (sycamore), triangular (alder), square (wahoo), pentagonal (chinquapin), hexagonal (buckeye), star-shaped (sweetgum), or variable.

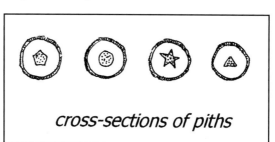

cross-sections of piths

A careful linear slice along a length of pith can reveal its inner make-up. If the pith is made up of continuous material, it is called "homogenous." Most trees possess homogenous piths. If a solid pith is interrupted by transverse bars or septa (like the tulip tree), the pith is called "diaphragmed." If the transverse bars separate empty spaces (picture a split cane of bamboo), the pith is said to be "chambered." Black walnut, sweetleaf, and pawpaw exhibit this pattern.

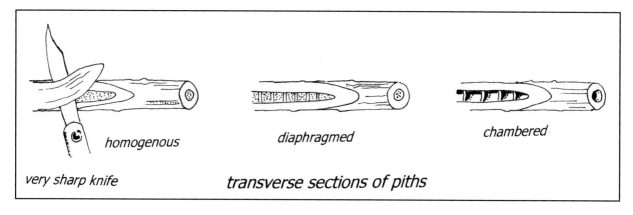

homogenous   *diaphragmed*   *chambered*

very sharp knife   *transverse sections of piths*

<u>**Using a winter key**</u> – Now that you are ready to start keying out winter trees, begin with a known plant. This will help you to get a grasp of a book's descriptions as they apply to the tree. Study the written description of that known tree until the words in the book jibe with the details you are seeing. Do this with as many known

trees as possible. If you can't begin to identify a single winter tree, go to a Nature trail that has trees labeled or ask a tree-savvy friend to go with you on an identification walk. As I mentioned in the first chapter on plants, I highly recommend that you first get to know your schoolyard or home area or neighboring woods. It is a place where you can frequently study the same plants through the seasons. However you do it, start with known quantities. Later, the first time you stand before an unknown tree and successfully key it out, you will feel an uncommon sense of pride. Here are a few to try in the early rounds: dogwood, maple, hickory, tulip tree, and sweetleaf.

Winter Remnants of Herbs

The so-called "weeds" often leave behind persistent remnants of their structures to aid in winter identification. For example, jimsonweed's (*Datura stramonium*) large spiny seedpods are unmistakable. Its cousin, horse nettle (*Solanum carolinense*), dangles a small yellow fruit (resembling a miniature tomato) from its thorny branches. (Both plants are toxic to ingest!) Square-stemmed mountain mints (*Pycnanthenum spp.*) have dry, aromatic flower remnants clustered densely in heads at the tips of opposite branches. Heal all (*Prunella vulgaris*) has a dense head of capsules like a rough bottlebrush. Indian pipe (*Monotropa uniflora*) is woody and erect with a small-mouth urn for a capsule. The urn looks like it has been grasped by a wooden flower at its base. Wild carrot (*Daucus carota*) grows an umbel of flowers that turns into a woody "bird's nest" holding bristly seeds. Dogbanes (*Apocynum spp.*) hold onto long slender pods, and their once cranberry-colored stems and branches darken to a purple-black.

One of the most handsome seed capsules that persists through winter is a wildflower called "seedbox" (*Ludwigia alternifolia*). Its seed capsule resembles a square tan cottage (just large enough for a ladybug or two) with a rounded roof, at the center of which is a "chimney" hole. Joe Pye weed (*Eutrochium spp.*) stands tall, its hollow stem still whorled with vestiges of petioles. Evening primrose (*Oenothera biennis*) sports inch-long, wooden, banana-shaped seedpods partially peeled from the top along four seams. Wild yam-root (*Dioscorea villosa*) holds its three-sided, heart-shaped, papery "Japanese lanterns." Cattails (*Typha spp.*) hold erect their fuzzy "hotdogs on a stick" (female flowers gone to seed). New Jersey tea (*Ceanothus americanus*), a small shrub, displays clusters of tiny white plates balanced upon tiny sticks. The yellowed stem of Solomon's seal (*Polygonatum spp.*) still shows evidence of its two-ranked alternate leaves and the fruit stalks forking down from the axils. Finding Solomon seal's desiccated wisp of a stem in a winter survival situation would be a monumental boost to morale and general welfare. A nutritious and delicious rhizome lies just underground.

Your best method of learning these skeletal remains of plants is to get to know the plants in their prime and mapping their locations so that you'll know where to find them out of season.

<u>*Keying Out Winter Plants*</u> – Now that you have used your winter key on

types of trees that you already knew, take that key to trees that you do not recognize. Keep a record showing where those trees grow so that in spring you can return to key them using their leaves, thereby confirming your winter assessments.

<u>*Filling in Your Plant Map*</u> – Now that you are well on your way in

the skill of identifying plants – both in their green and in their dormant seasons – update the map you made in Chapter 4 (<u>*Mapping Your Land*</u>) by labeling the plants you have identified. Post your map in a visible place and watch your child's (student's) interest in plants build. (Young students love maps of places that they know.) Add to your map every new discovery, including new specimens you transplant into the area. Send students on a quest to find a plant marked on the map and to verify its name by using a botanical key.

PART 2

The Lore of Survival

"If you get lost out there, the world around you may seem your enemy, but it's not. It's just that you've forgotten what your ancestors knew a long time ago."

~ Natalie Tudachi, *Blue Panther Woman of the Anigilogi clan*, Let Their Tears Drown Them

Author's Note

The physicality of satisfying our essential needs today is virtually effortless – like strolling through the aisles of a grocery store or turning a knob to heat an oven. So why turn back history and roam forest and field to dig up edible roots? Why twirl a stick in your hands to near exhaustion in the quest for fire? Isn't all that ancient lore outdated?

It might seem so, until some part of the current system fails. When a taken-for-granted service of our modern lives is not available – like electricity or gasoline – the "old ways" become invaluable. In such scenarios we are forced to look back at the ingenuity of our ancestors who knew how to recognize and utilize the resources of the natural environment.

Whether seashore or woodland, prairie or mountain, desert or tundra, every different landscape demanded competent survival skills of the first people to settle it. Each discrete land-type required a specialized lore, because humans had to meet their needs not with resources they *wished they had*, but by using *what was available*.

To one man, the building block of his home was snow or ice. To another it was grass or mud or animal skins. Some utilized logs. Others used sticks.

Here in the forests of these Appalachian Mountains where I live, caves were certainly appropriated for shelter, but of course caves could not be moved to the most favorable settings. (One etymological interpretation of the word "Cherokee" translates as "people from the caves.") In the earliest times, new tenants of the land simply settled for what they could find.

When early inhabitants of southern Appalachia began to look at choicer locations (and less dank living quarters), their first homes were comprised of *sticks* and *mud* and *leaves* – the components of a wattle and daub hut with a thatched grass roof. These building materials were (and still are) found in abundance in the Great Eastern Woodlands.

> One way to look at the history of humans is to follow the "evolution of comfort."

Survival skills allowed paleo-settlers to *get started* in the process of *staying put*. In time, grander schemes replaced the raw hardscrabble tasks that were at first required. For example, a one-room hovel might have been expanded to a more complex home by adding the space needed to perform a variety of chores. Now there was room to store water in the home, because free time had opened up for the making of vessels composed of clay dug from creek banks. This addition of crockery eliminated the constant going to and coming from a stream. In other words, a resource (water) had been harnessed, so to speak, by bringing it inside within easy reach.

It was only natural that humans would strive to invent less demanding solutions to their basic problems. We're still doing it today. In fact, one way to look at the history of humans is to follow the "evolution of comfort." In the story of water, anthropologists consider it a pivotal moment when humans stopped lowering their faces to a stream and learned to cup water in their hands and raise it to their mouths – a posture better suited to staying alert and alive by seeing what predator or potential enemy might be approaching.

Consider where we are now in the story: channeling water through the walls of our homes by way of pipes and then again out of our homes once we have made use of the water. What yet-to-be-discovered technique might the future hold for making water available inside our homes?

Regardless, we are now enjoying a very comfortable juncture in the story. We rise from our chair, take a few steps into a nearby room, turn a handle and … water flows. We drink. The primitive approach in a modern-day survival situation would include building a fire of dead hardwood and letting the wood burn down to coals, burning out a concavity in a chunk of wood by using hot coals and a blow-tube (a hollow plant stem like Joe Pye Weed or rivercane) to superheat the coals. Next, scraping the burned-out depression with stone and sand to produce a finished wooden bowl, filling the bowl with impure water from a stream, heating bird's-egg-sized stones on the bed of hot coals and at intervals dropping red-hot stones into the bowl of water to bring it to a boil to purify the water.

At my wilderness school I teach students of all ages about Nature. By far, my most popular classes are those that explore how the Native Americans perfected their survival skills – lore that tied them so closely to the natural world that the results of that partnership went far beyond the pragmatic. Along with utilitarian appreciation came respect, reverence, and gratitude. A relationship formed between human and Nature. It could be no other way.

Whenever I ask my students why they sign up for such a course, I hear lots of perspectives on the relevance of survival skills for our time. These rationales usually boil down to two basic categories: autonomy and enriching any outdoor experience by using and integrating with natural resources found in the wild. But I expect a third category lies deep in the subconscious … in a dimly lit room called "atavism."

Might you *really* ever find yourself in a life or death scenario … you against the elements? Thirty years ago, I prefaced my survival skills classes with this statement: "Chances are, you'll never find yourself in a true survival situation … like walking away from a plane crash in the Andes or being car-jacked in the jungles of Central America and left on your own." That's what I once told my students. Not now.

Of course, the more adventurous you are, the more your odds increase in coming face to face with a survival challenge. After all, not everyone flies over the Andes or travels jungle roads. For that matter, it's only a small percentage of Americans who venture out on the trail for days at a time with packs on their backs. But the times have changed to make us all logical candidates for learning these ancient skills.

Three Reasons to Learn Survival Skills

We live in a different world now because of the escalation of terrorism. There are people in other countries who feel that America is long overdue on being at the receiving end of foreign attack and devastation. (Pearl Harbor seems not to count, because of its distance from our continental shores.) We, who engineered the horrors of Hiroshima and Nagasaki and head the list of the planet's most voracious consumers, present an image to many that is spoiled and arrogant. From the lessons of September 11, 2001, we have a new awareness that – within the scope of terrorism – anything is possible. Almost certainly, there is more tragedy to come.

Short of nuclear holocaust, an act of terrorism could demand from us at least a few survival skills. Consider a deliberate contamination of a municipal water source. Water from surface streams (though now almost nationally polluted) would come full circle and once again be a primary resource just as it once had been. Knowing how to render such water drinkable would be imperative. Drinking that stream water "as is" would be a mistake – one that could level you with a sickness so severe as to incapacitate you. Racked with pain, you would become, paradoxically, dehydrated and too miserable to perform the tasks necessary to stay alive.

What about an extended power blackout? In winter, if you don't have a wood heater or fireplace, you'll be setting up a grill or permanent fire pit outside your home. You'll have to haul water to it – that, or make your fire pit near the stream. You already own all the containers and cookware you would ever need for collecting and boiling water.

If our system of commerce and the transporting of goods were brought to a halt, grocery stores would empty in just a few days. After depleting our pantries, we would be forced to venture outside for our food. Hunting and fishing would enjoy the revered status it once held in pioneer days. Those good at it would be renowned and in demand, but the forests and streams would be overwhelmed by sheer numbers of people trying their hands at it.

As for our other needs, virtually all of us possess at this moment a lifetime supply of shelter, clothing, and tools. But how much do we know about successful gardening? Seed storage? Or foraging for wild plants? The experienced farmer and the naturalist would become our mentors.

There is still enough wild land in America to challenge a person who is unexpectedly stranded. Every year, we hear of someone's unplanned ordeal (or demise) in wilderness – the person who wandered off from the group, the traveler who ventured off on an unfamiliar route and ran out of gas in a remote area, the solitary adventurer who pushed his/her limits.

There are probably places in your own county that would qualify as "wild" enough to pose a problem to you if you weren't prepared for an unexpected stay overnight. If this were a winter misstep, you would not likely succumb to starvation or thirst, but to hypothermia. Losing core temperature is man's greatest survival nemesis in the cold seasons – especially in wet weather. (Even in summer, a cool mountain night can prove deadly to a wet person.) Hypothermia occurs when the

> *Virtually all of us possess – at this moment – a lifetime supply of shelter, clothing, and tools.*

loss of body heat reaches the point where endothermic heat cannot be retrieved by natural body metabolism. Without help afforded to the victim from an external heat source or a hot drink, this condition leads almost certainly to death.

If you think you might be exempt from the threat of hypothermia because you are physically fit and plan to fend off the deadly cooling process by running or some other strenuous exercise … think again. The debilitating stage of hypothermia slips in stealthily and soon after you had been feeling capable of taking care of yourself. It denies muscles the ability to function. Before you realize your condition, the option to exercise is gone. (I write this from the experience of a youthful misstep … and with gratitude to the kind stranger who chanced to find me.)

A most useful skill would be knowing how to insulate your clothing with materials from Nature. And how to get a fire going. If your stalled car is nearby, you would have a tremendous asset in its shelter from wind and rain. But people have frozen to death inside their cars, where building a fire is not a smart option.

You might even have matches or a lighter for an outside fire. But I have learned through my classes that, even with those incendiary assets of technology, most adults lack sufficient fire-making skills – especially in wet weather. (Did you know that a forest that has been soaked for days in a driving rain still has much more dry dead wood that wet dead wood? Would you know where to find it? This is covered in *Secrets of The Forest, Volume 2*.)

The last "survival rationale" is my favorite, because it is for anyone who simply yearns to be "out there" in God's world. Survival skills integrate you with the land like nothing else can. Think of taking off on a hike or camping trip and purposely leaving matches at home. Or leaving behind food or tent or sleeping bag. Your surroundings become the source of some of your necessities. The forest reveals itself as the bearer of gifts, just as it did for the native people who preceded us. How differently your eyes would look upon each gift as you learn to revere the simplest parts of your environment – things like sticks, leaves, stones, vines, roots, and bark.

Survival knowledge lightens your load and sharpens your eye. It brings immediacy and purpose to the details of the forest. And because you use these items to your advantage and comfort and sustenance, you earn a sense of connection to the forest that will last a lifetime. You will be not simply *in* the forest … but a working part *of* it.

I live on what was once Cherokee land, so it makes sense to me to learn what they once knew to a man, woman, or child. If I discover something that at first might seem like my innovation, I know it is really nothing "new." A Cherokee learned the trick long before my time. Therefore, as my relationship with the forest has grown, so grows my bond with these first people, who understood the value of each piece of the puzzle of the place that we should call "the real world."

"Right now we ain't got nothin' prepared. We're a long ways from anybody, and we're hungry. We'll be working hard for a few days. But as we start collecting what we need and get settled in, our situation won't seem so desperate. Now ... help me with that log. We gotta get a roof up over our heads."

~ *Lonny Deerfoot,* Requiem to the Silent Stars

CHAPTER 12
The First Step
~ *Getting Started in Survival Skills* ~

Only two per-cent of the adult students whom I host carry a blanket in their cars. Even less carry water, matches, extra clothing, or food supplies. These are not fanatical considerations, but practical ones. Perhaps the world's congested population suggests that these safeguards are not necessary; that is, there will always be someone close by to help us. This frame of mind lulls us into a false security, even when we enter vast areas of low population density or wilderness that, thankfully, still exist.

Among my hard-core whitewater friends, with whom I have plunged into adventures made more "on the edge" by the remoteness of the journey, I was consistently met with one baffling fact on day trips: I was usually the only one in the group carrying emergency supplies. I am not referring merely to a first-aid kit but things like dry matches, a bundle of resin-rich dry kindling, a wool sweater, and a few overnight needs. Complacency can be found anywhere.

There is a strong and convenient propensity for humans to plan according to *how they would like things to go* ... rather than *how things might go*. Simply imagine what *could* go wrong. And if that happens ... what would you need?

Even if primitive fire-making is your strong point, if someone falls prey to hypothermia, you'll want a fire blazing within minutes ... not in half an hour. Matches are invaluable. My journeys down rarely explored streams coincided with high water levels, which often meant rainy-day excursions. Not only did such weather increase the chances for hypothermia, but also it slowed down fire-making time. I

considered a supply of dry matches and highly flammable lighter-wood a life-saver, and in those forty-five years when I was an active canoeist … on one occasion it was.

It was on a twenty-mile-long, remote section of river that had been planned as a day-trip in December, but one of our party had misjudged his stamina. With our group of three only as fast as its slowest member, the moonless dark caught us with seven miles to go. The difficulty of the rapids made pushing on blindly impossible. As ice accumulated at the river's edge that sub-freezing night, my companions slept toasty-warm by a fire made in minutes in pitch dark. Very high on the list of assets that night were: a flashlight, wax-dipped matches, splinters of lighter-wood, and extra sweaters. That one overnight (out of a lifetime of trips) justified my precautions. It could have been our last trip.

Survival materials kept in a car, in a canoe, or in a living abode are simply evidence of common sense. One has only to think about what might be needed in case of a highway breakdown, running out of daylight on the river, or a power outage for a given season. At home the degree to which a person prepares for potential disaster can vary from caching bottled water and canned foods to the hoarding of firearms and ammunition.

Wilderness survival can be approached on different levels of technology, too. There are fire-starting devices, water-purifying pills, space blankets, special heat-conserving clothing, chemical hand-warmers, nutritional tablets, compact multi-use tools, portable communications devices, GPS, weaponry, and more. This book will deal with the primitive techniques used by the Cherokee, who knew these Appalachian wilds as no people will ever again know them.

These skills can never become obsolete; because, whereas the availability of purifying tablets, synthetic materials, processed chemicals, ammunition, and machinery could conceivably come to an end … the gifts of wood, leaves, stone, earth, and animal, hopefully, will not.

There are wilderness programs that take students who want to "survive" for a given duration under the tutelage of a staff. Though such an experience is not without its merit and lessons, it might teach more tolerance than proficiency. At my school I have hosted many students who had attended such courses. They assured me right at the start that they were old hands at all these skills and had just come to Medicine Bow to "brush up."

As it turned out, they were not skilled. My sense is that a student might walk away from the trying ordeal of a full-bore survival program thinking he knows how to "survive," having put up with a week or two of discomfort. Such an attitude represents one of my least favorite American aphorisms: "Been there, done that."

I suspect that one appealing aspect of undergoing such an "ordeal program" is that it is supervised, and the student feels that cushion of the removal of risk of life. This is a sound consideration. No one should enter into a voluntary survival challenge with death as an option. But in reality these students were skimming the surface of survival … just as they are at my school. The difference is that, at my school, they know it. At Medicine Bow they'll put a lot of sweat and effort into

each hands-on project, but more importantly they will walk away with copious notes that guide them toward mastering each project *on their own*. It takes years.

Am I criticizing such survival programs? Only if they purport to make a "survival expert" out of a student in a short time. Taken as a source of lessons, there are some excellent teachers offering these programs all over the country. And learning about tolerance does have its merit. Furthermore, a forced ordeal does have a valid place in triggering *action* – especially as a therapy for certain behavior-rehabilitation groups (at-risk youth) that need lessons in accountability and discipline.

My philosophy embraces another route: a self-motivated series of "partial ordeals" that accumulate over time eventually to comprise the full-scale, voluntary survival trip. Each step of this endeavor is preceded by an earnest quest to master the one skill that will be tested in the outing. In other words, after the training, after learning from a teacher (a person, a book), and after lots of solo practice, the student goes on an outing by his own deliberation and determination, willing to take full responsibility for successes and failures, but with a back-up plan for failures.

That's the way we shall approach the skills within these pages. If you are motivated more than the average student, go out on your own and see what you can figure out about shelter or fire-making or building traps. **But don't let this experimental method carry over into your use of plants as foods or medicines. Never make assumptions about unknown plants. One other area that should not be approached experimentally is hunting. It is unfair to the wild animals. Every hunter has an obligation first to become unerring with his weapon.**

The number-one asset a person could possess in a true survival situation is **"sangfroid"** – the ability to remain cool and collected while under pressure. An unfrazzled presence of mind is essential to respond logically to needs in a stressful situation. I suspect a person either has this or doesn't, but I do believe one can improve in this area by sheer will power and by exposure to challenging situations. It is a *choice*, and *choice* is, by my estimation, one of the great human characteristics ... *to be a way we want to be.*

If *tolerating* a week of survival class helps a person realize his/her level of (or lack of) *sangfroid* and to consciously work on the will to be steady and productive, then that survival class experience becomes more important. And if such an ordeal spurs a student to return home and practice those learned physical skills toward mastery, even better.

But I doubt this is the outcome for most ... unless a student enters into the experience with a predisposed commitment to and serious hunger for such lore. My complaint lies more with the impatience of the student who wants to "own" survival skills as he might own a car: Buy it, step in, crank it up, and do the best you can out on the highway. Survival skills can't be mastered in days or weeks or even months. You can't buy them. You earn them.

A survival student should consider himself a newborn in a primitive tribe and let the years accrue with earnest endeavors toward mastering these skills. The journey can be more fulfilling and comfortable with patience and planning.

The Pace of Learning Survival

I suggest you approach the survival skills within these pages the same way my students do in classes. Immerse yourself into an individual hands-on skill, but do not expect to master it right away. **Survival methodologies can be *learned* from a teacher or book, but proficiency must be *earned* by self-motivated practice.** Even better … survival skills can be learned by trial and error experimentation, failing, trying again, failing, and *then* consulting a book or teacher, trying again, succeeding, and finally *owning* the skill.

I place a high premium on solo work. It is the mark of true dedication to be in the woods *alone* working on a project. Your focus on the work becomes absolute. Consider that as you train.

I encourage my students to take the lessons they have learned and incorporate them one at a time into a camping trip. For example, they might pack for a trip as usual, with one exception: *no matches*. The journey need not be long. I favor setting up one base camp for the entire camping trip, so that traveling does not become the journey. The journey is the wandering, the searching, the gathering, and the preparing of all necessities. The journey is the work put into achieving a flame and the respect afforded that fire so that it does not die.

On the next trip, that student might omit tent or tarp. All other needs are carried, including matches. Next trip, leave the sleeping bag at home but take everything else. Next trip, no sleeping bag, tent, or matches. Next, take everything except food. And so on.

My survival classes are not set up as ordeals … but they *can* be if a student so chooses. To my students I offer the ways that they can make the class more challenging for themselves. For example, I assign several students to work as a team to make <u>one</u> winter shelter (designed for one body). Each team member has the option of time-sharing through the night to experience the efficiency of the group's work. Often only one student is interested in the overnight adventure, in which case he/she has the sleeping rights in the shelter for the whole night.

Thirty years ago it was standard practice for me to ask every student to build a solo shelter. That night each tried out his abode with a nearby tent and sleeping bag as back-up. Today it is rare to find a single student willing to work solo on the project, much less spend the night inside his handiwork. Of course, any teacher would look favorably upon an industrious one willing to meet the challenge; after all, self-motivation is what my survival philosophy is all about. But, the times have changed; and, for whatever reason, there are students who cannot or should not undertake the long night in a shelter. Their reasons are personal and not mine to question. Even if they never enter the shelter, I know that they leave my school with the construction knowledge that will allow them to enter the challenge at the time of their choosing. It is a task better met, I believe, if it is the student's idea.

The early phase of a survival scenario can seem overwhelming. There are so many implements to gather, so many considerations for which to plan. A person thrown unexpectedly into this situation can become frenetic. Taking projects one

at a time, without the frenzy, is infinitely more enjoyable and instructive. Learning survival skills as *integrating with* rather than *battling against* Nature has many advantages. Going at your own pace – rather than the pace of trying to stay alive – makes the learning something you want to return to.

Being an accomplished survivalist is like entering an art museum and seeing not only the paintings on the walls but also knowing the location of everything else in the building: carpet seams, electrical outlets, light fixtures, water fountains, bathrooms, window panes, trashcans, staples in the printed programs, overhead wires, floor tiles, etc. The eye learns to be on the lookout for all needs at all times. It is difficult to jump into this frame of mind early in survival education. It can develop naturally as skills are accrued one at a time.

As a whole, survival practice elevates your camping trips to a new plane – part of which is physical and part spiritual. Gratitude soars to new heights. To look longingly at a meal of tubers, wild greens, and larvae cooking on a bed of coals might be the ultimate prayer of thanksgiving.

This gratitude travels home with you, too. When you next sit down to a plate of prepared food in your home, you may feel something more appreciative and spiritual than you have ever before experienced with home-cooked food. Survival skills bring you closer to the real world, aligning your spirit with the living energy of the wild. Paradoxically, survival skills also heighten your appreciation for the inventions of our manufactured world.

After returning home from a winter survival trip in which you stayed in your first hand-made shelter, you might stare in awe at a carpenter-hung door, its hinges, and its snug fit inside its perfect frame. You will be one of the few who understand the beauty of a perfect door. The memory of that cold draft around your head in your shelter will inspire an improved primitive "door" on your next shelter in the wild.

Survival is complex and multi-dimensional. Eventually, the survival explorer will amass all the skills and learn to look for everything at once: logical sites to set traps, sources of tinder, best shelter location, plant foods, dead wood, rotten wood, cordage fibers, fire-kit components, fruits and nuts, useful rocks, digging stick, etc. Imagine the level of observation skills you will have earned at this stage. Think of the scope of such a comprehensive and condensed survey of your surroundings. Imagine the self-esteem gleaned from such self-discipline and autonomy. This is the real world. Knowing it … looking at it with such precision … lifts a person to the higher plane of appreciation that was once the way of the people who originally inhabited "our land."

> *To look longingly at a meal of tubers, wild greens, and larvae as they cook on a bed of coals might be the ultimate prayer of thanksgiving.*

"The dress of the women consists of a robe ... confined across the breast with a string ... the most esteemed and valuable of these robes are made of strips of the skins of the Sea Otter net together with the bark of the white cedar or silk-grass."

~ *Meriwether Lewis*, The Definitive Journals of Lewis and Clark (*Univ. of Nebraska Press*)

CHAPTER 13
The Ties That Bind
~ Cordage ~

On my self-imposed solo survival trips through the years – experiences I have willingly chosen to enter for keeping my skills well-honed and for maintaining that "edge" for teaching *and* for the pure adventure of it – I have always considered the first day to be a crucial one. There is a lot to be done. Rarely did I make this a "migratory" trip. Instead, I set up a base camp to use for the duration of the outing. The first order of events was always to get a shelter built in case of rain.

Such work calls for a balance between steady dedicated labor and not getting over-heated, because with work comes sweat. Sleeping in damp clothes or wearing even a thin coat of sweat on the skin makes for a *very* uncomfortable winter night. Evaporating moisture robs the body of precious heat. Even in summer, shivering through a night in the mountains can leave you physically and psychologically spent on the following day. (For this reason, my last act before crawling into a leafy bed has always been a stream-bath followed by a fire-dry.)

After the main structure of the shelter frame is complete, the most work-intensive aspect of shelter-building begins. In a deciduous forest the task is to gather dead leaves off the ground and haul them to the shelter. These leaves are a primary material of insulation and water repellency. (You'll learn all about the construction of this shelter in Chapter 14.)

In my earliest survival trips, I crouched and scuttled backwards as I raked with my hands to pile up leaves. Then I carried the leaves in my arms to my shelter frame to arrange them properly. No matter how mindful I tried to be in minimizing the perspiration factor, working up a sweat seemed unavoidable – even on the

coldest of winter days. The problem was that on the first day of my trip, I didn't have the luxury of performing the job leisurely. In "survival mode" (an inspired, energetic frame of physical work) it took me four hours to complete a dependable winter shelter. That's a big chunk of the daylight hours in which there are so many other jobs to address … like fire and water and food.

Early in the history of my survival trips, out of sheer need, I solved the problem with a tool that cut my shelter-building time almost in half! I re-invented **the rake**. With my new tool I was no longer forced to lean over and work in a squat, and I could sweep piles together much faster. The effort of leaf gathering was so greatly reduced that I could work nearly sweat-free. All that was needed to make that rake were three select sticks and five short pieces of *rope*.

Rope, as far as history books are concerned, is one of the most underrated inventions of humankind. There are a few materials available in the Appalachian forests that can serve as ready-to-use cordage, but quality rope-making requires clustering together and twisting plant fibers into a reliable weave. No matter what it was made of, rope was profoundly important to the development of civilization. Rope bound together the parts of an early tool. It lashed together the structural components of a home. If rope makes a home … and if homes make a village … then it follows that rope is the binding of civilization. You can appreciate this when you consider that the substitute for cordage in modern construction is the carpenter's nail. Like rope, nails hold things together. Consider for a moment how many nails are sunk into the wood frame of your house.

And consider this: very slender rope (thread) is the unit of all woven material. With a little know-how, if you can make cordage, you can make a crude blanket or clothing or a sleeping mat.

Whenever I visit a school classroom to present a Native American program, I take along a section of bark from a dead tulip magnolia tree. I bring this particular bark because it is the most abundant source of cordage fibers in my area. I know that the students will be able to find tulip trees around their homes, and when they do – if they try making rope on their own – this might be the spark that ignites their interest in Nature.

To initiate a conversation about rope with children, I usually employ the classic survival situation: getting lost in wilderness. We try to imagine ourselves stranded somewhere. There are always some who have read Gary Paulsen's *Hatchet* or Jean Craighead George's *My Side of the Mountain*. Most have seen "reality TV" survival shows.

"What if you're lost," I ask the children, "and need string for a trap to catch food. How could you make some string?"

Invariably hands go up, and someone always answers, "I would use a vine!"

Because I place a very high value on accountability, if we are outside I point at that child and say, "Let's go to the woods! I want to see how you do this."

As it turns out, few vines are immediately usable as a rope. As a simple test, tie a section of vine in an overhand knot – the same knot by which we begin tying our shoelaces. When you pull the knot tight, almost every vine will break.

Failures are wonderful. They teach. They place children in a frame of account-ability where the students realize that rewards are not going to be automatic. They have to be earned. Failures force children to experiment more, to forge ahead. Fail-ures take them from the manipulated illusions of movies and video games and set them down in the sensible terrain of reality. Failures ground all of us. They keep us from fooling ourselves about the nature of the world. They get us involved.

This impromptu quest for rope is time well-spent. After a few experiments by eager children, it's time for a lesson on how to make rope. I pull out my tulip tree bark (or better yet, we find some on the school grounds) and soak it to make it more pliable.

If my students are 8-years-old or more I like to perform the rope-making les-son in silence. I ask them simply to watch. I instruct them to remember every twist and turn that my hands make and to memorize which hand (right or left) performs a specific twist or turn. Because the technique is so repetitive, it is not asking too much of them to absorb the technique without words.

There will always be a few who fail to make a successful piece of rope. This provides a grand opportunity for some students to tutor others. Few things are more gratifying to me than to watch these new "young teachers" at work.

I prefer the silent lesson, because it simulates the way Native American chil-dren learned the craft – by simply watching, by absorbing everyday life. Elders welcomed children to spend time with them around the village. If adults were gathered to work on a craft or to talk, children could join them and glean tidbits of education.

In my school visits, I tell the students that we are replicating this historic learning experience. When I begin to move the fibers through the rope-making formula, the students are 100% in the present, their eyes glued to my hands and to the beige fibers that I hold. Unknown to them, they are reliving the past.

When my visit is over, I leave the group believing that the day's lesson will not go down as a "flash-in-the-pan" bit of entertainment – something enjoyed and then forgotten. I wave goodbye and see hands come up to return the gesture, their wrists adorned by a bark bracelet. I know that they will wear that primitive jewel-ry for a long time. Perhaps they'll find bark elsewhere and teach someone else to perform this ancient craft. And maybe someday, when the need arises, they'll *use* such a rope.

<u>The Ties That Bind</u> (and the ones that don't)

<u>The Ties That Bind</u> (and the ones that don't) – With your child or class-room of students, set the goal of simply finding in Nature something "natural" that can be tied into an overhand knot and pulled snug without snapping. Search in a forest, field, or vacant lot. Encourage experimenting with plants, but first be certain that your wards know which plants not to handle. Survey the area together and discuss ground rules for safety.

Then use a 20 lb. barbell and challenge your class to make a natural rope that will suspend the weight from a tree limb without breaking the rope.

ROOTS

First, let us consider tree roots. These evergreens – **pines (including hemlocks, firs, and spruces)** – produce rootlets that are both tough and pliable enough to use as rope without any preparation other than digging them up. If a rootlet does not pass the simple "overhand-knot test," slit the root lengthwise on one side and strip out the woody tissue at its core so that you are using only the root rind. Add to your tree root list: **black locust, mulberry, juniper, elm, walnut, hickory, and willow**. Use a sharpened digging stick to loosen the soil as you follow a rootlet to the desired length. Depending on tree size, these roots might be rather short sections – say, 4 feet or less from a small sapling. But 4 feet is plenty of rope for most tying or lashing jobs, and root rope is usable immediately.

Ropes From Roots
– Seek out a place where very young pines grow in abundance. These trees should be 5' to 8' tall. With a digging stick, churn up the dirt just inside the drip-line and locate a rootlet ″ – ¼″ in diameter. Follow it by digging a trough, being careful not to gouge into the root itself. Use your fingers to work your way along its length and expose the longest section possible. Sever it from the root system and test it by lashing together two sticks.

WOOD FIBERS

On rare occasions **rotting <u>wood</u> fibers** can be found for making cordage. To find cordage fibers it's simply a matter of being curious and willing to experiment. **Green branches** of Alabama fringe tree (*Chionanthus virginicus*) and leatherwood (*Durca palustris*) can sometimes serve.

INNER BARK FIBERS

By slitting a hickory branch lengthwise and pulling its bark away from the woody branch, another "instant rope" (the bark) is in hand. If the bark is too stubborn to be stripped off a branch, pound it with a smooth, heavy stick on a smooth work surface (a dead, bark-less log). Other trees can be used this way, too. Experiment and find them. Your successes may vary at different times of the year, because after the growing season inner bark eventually adds layers to the tree's sapwood and to its outer bark.

Inner bark fibers of certain trees provide a plentiful supply of rope materials. Some fibers are usable only after the tree is dead, others only when green. Some can be used either way.

The most abundant and available source of dead inner bark cordage fibers in Southern Appalachia comes from the **tulip magnolia**. If you were to cut a strip of green inner bark from a living tulip tree and try to use it as rope, the bark would likely break even as you pull it from the tree. But find a *dead* tulip tree – especially one in a moist valley – and peel off its bark (outer and inner together) and if the timing is right you might encounter beautiful laminae of beige inner bark that strip off in ribbon-like layers. Finding the tree at just the right stage during the tree's decomposition is somewhat serendipitous. Recent weather conditions are one factor. Shade is another. But even if the bark is too stiff or brittle to be used right

away, soaking it in a creek can render it useful. The duration of soaking is variable, depending upon the bark's condition. Once your eye has learned to recognize the tree's appearance in its prime rope-making time, you will encounter usable material regularly.

A standing tulip tree will often grab your attention with its "pennants" of sun-bleached inner bark dangling from dead branches still attached to the trunk. These flags are a sign of activity by animals that have a keen interest in the bark for nest and bedding materials. Squirrels, mice, and birds strip off convenient pieces from the tops of branches but leave the harder-to-get-to layers on the underside.

Ideal dead tulip tree bark results from wet and dry cycles. A tree lying on the ground is more likely to have its lower side receive the necessary wetness by soaking up moisture from the earth. On the other hand a horizontal fallen tree in *constantly* wet ground might have its upper side develop better bark for cordage while the bottom side rots. In the photo below notice how dead, brittle tulip tree wood often breaks not into 2 but 3 pieces. The long, 3"-wide slab of tan material hanging from the stick is inner bark (still attached to its outer bark on the other side).

My favorite hunting ground for bark is a dense grove of young tulip trees standing in a moist valley, most of the trees being wrist-thick or smaller. Some of these trees do not survive the competition of overpopulation. They die but remain standing. In winter I spot them by their lack of vitality in color or by fissures in the outer bark, and then I double-check by shaking the trunk to listen for the dry, trebly rattle of dead, resonant wood.

Pull down the tree and break it at different points on the trunk to check the condition of its inner bark. Many variations are to be found. Some bark is so old, splintered, and brittle that it will be futile to work. Some may be too recently deceased (still too "green") with its inner bark firmly attached to the outer-bark, which might cause it to tear too easily. This bark will need some aging/soaking before it is ready for rope-making. Others may be too stiff but holding together in thick ribbons.

tulip stick broken

These fibers can be manipulated through wet-dry cycles at a creek over the next days or weeks or months. Submerge and anchor the broken-up tree for a few days. Remove for a few days to let dry. Repeat as needed until the natural glues dissolve to allow the separation of inner bark layers.

The real treasure can be found near rivers, where fallen trees on the stream bank go through their wet-dry cycles by the fluctuation of the water level and by exposure to the moist night air of a valley. On many occasions I have gathered over a hundred pounds of ready-to-use fibers from a single dead tree near a river.

If sizeable ribbons of bark are not available, short or thin strands can be put together side by side, overlapped for length (just like wool from sheep is carded) and spun into loose strings before weaving. <u>Dead</u> branches from living trees can provide good ribbons of dead inner bark.

Soon enough – if you actively search – you will find a perfect dead tree with long strips of bark. If you discover a dead, too-thick-to-fell tulip tree still standing in a valley, you might circumcise the bark near the ground and then pull off long strips as you step back from the trunk.

Locating a Source of Rope Fibers – Walk the wooded floodplains and

bottom lands to find a grove of young tulip trees. In summer, identification is easy. The broad leaf of tulip magnolia terminates in a notch (a shallow V) rather than a point. Break <u>dead</u> trunks and branches until you find beige, ribbon-like inner bark fairly bursting from the tree. If stiff inner bark has not yet begun to delaminate into ribbons, submerge/anchor harvested trees in a creek, let dry, and repeat as needed.

It is best to use ribbons that maintain a constant width along the length. When using uneven ribbons, add new fibers to narrow spots or shred fibers away from wider spots. Soak before using.

Making the Double-Helix Rope

Hold a ribbon of inner bark between the thumbs and index fingers of both hands, letting 1½" of ribbon show like a bridge between thumbs. Twist the right hand forward (rolling the ribbon between thumb and index) and twist the left hand backward (rolling again). Shift your grip so you can repeat this motion without losing the twist just achieved. When you can feel significant twist-tension, pinch the ribbon tightly as you bring your thumbs together. The bridge will automatically form a loop. Because this dead bark is non-toxic, you can use a tooth to anchor the loop in place, while your hands continue to hold their respective strands.

For the rest of the rope-making process, keep the two strands taut by pulling against your tooth-grip. Keep the two strands angled like a V – not pulled to the sides like a propeller, not stretched out in front like the arms of a platform diver, but halfway between these angles. This puts about 90° between strands at the point of their juncture at your tooth.

Repeatedly twist the right strand forward and the left backward until you feel twist-tension again, then carry the left strand over the right and switch strands in your hands. During the switch keep the strands taut. Twist … crossover … switch. That's the pattern, over and over, until you've run out of bark. To tie off this rope so that it cannot ravel, place the two leftover ends of the strands together, treating them now as a single piece. Tie a simple loose overhand knot anywhere in the rope and before tightening the knot coax it to the spot where you want

the knot to be fixed. What makes this rope most strong is the twisting. (It's harder to break fibers that cannot be stretched out straight.) Consider all the twisting that goes into this craft: each strand is individually twisted, but so are the two strands as they intertwine, making the rope a double helix. To ascertain if you have done a proper job twisting, look at an **individual hair-like** <u>fiber</u> in your finished rope. If its linear direction is the same as the rope's, you have done well.

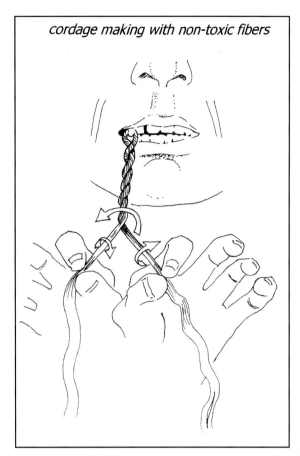

cordage making with non-toxic fibers

At any time during the making of the rope, you can release it from your mouth and, if you have performed the twisting and crossover correctly, the rope will not make a move. But if you twisted the wrong way or crossed over with the wrong hand, the rope will unwind before your eyes. This undoing of the work – as we say in my classes with youth – is the rope laughing at you. (If you perform all directional instructions backward, the rope will come out fine.)

<u>Making Rope to Wear</u>

– A bracelet or necklace makes a fine first project. The style of rope-making just described provides its own clasp. After completing an ample length of rope, untwist the loop end a turn and pass the knotted end through the loop.

<u>A Rope With Meaning</u>

– Make a bracelet for your child and clasp it around her <u>left</u> wrist as a gift. Ask her to do the same for you. The bracelet will serve as a symbol for some promise or commitment or quest to improve oneself in some manner. For example, if you feel that you don't spend enough time with your child … or if you think that you don't pay enough attention to what matters to her … share *not* the negative aspect of your topic with her but make a commitment to improve on whatever topic you have chosen. ("I want to come to more of your school performances.") Ask her to watch for that change. When she sees it, she should come to you and in a silent ceremony transfer your bracelet to your <u>right</u> wrist.

Ask her to commit to a similar quest concerning you. When you acknowledge her improvement, quietly transfer her bracelet. It is a poignant moment and each member of the bargain will benefit … as will the bond between the two of you. Rope can bind people, too.

Taking the Rough Out of Rope – Since most ribbons of bark are not perfectly smooth, a rope necklace or bracelet might feel scratchy to the skin. Look closely at your rope and you will see lots of little "whiskers." These can be easily and quickly <u>eliminated</u>.

Build a small fire in a safe place. When a steady flame has been established, attach your rope to the prongs of a Y-shaped stick. Slide the loop of your rope over one prong of the Y and weave the rope loosely back and forth over both prongs in a flat plane. Using a steady, gliding motion pass the rope horizontally through the flame several times to singe the whiskers. Flip the stick over and repeat, being sure to keep the mass of rope horizontal. When you no longer see any whiskers flare up like orange-red sparks, you're done. Simply remove the rope from the prongs and floss firmly between thumb and index to break away the scorched whiskers. Never let the rope hang down vertically in the flame. If the rope makes its own flame, immediately step on it.

Other dead trees that provide rope fibers are **basswood, willow, walnut, black locust, some ash, maple, oak, mimosa, and cherry.** Expect to find varying conditions of bark. Experiment with other trees as you encounter them. Discover for yourself what works and what doesn't. Don't rule out a tree from one experience. You may one day find its bark in a different state of decomposition.

For trees that offer rope from **green inner bark**, cut two parallel lines along a branch. Before stripping it off the tree branch, slice away the outer-bark, using the tree branch as your workbench. (Separating the two layers *after* stripping presents a safety problem in knife-use.)

"Green-rope" trees include **hickory, mulberry, hawthorn, willow, pawpaw, Osage orange, cedar, cypress, and basswood.** These strips need only twisting. Their strength is impressive. The downside is its slickness and seasonal availability. The autumnal/winter transition of inner bark to sapwood makes harvesting sometimes impossible.

Making Green Bark Rope – In a hardwood forest in early summer find a hickory tree with a living limb within reach. Make a cut 2 feet long on its upper side and then another parallel to the first, ¼" away. Flatten your knife blade on the limb and carefully shave away the gray outer-bark between the two incisions to reveal the yellowish inner bark. Sever the ends of the strip and peel the inner bark off the branch. Simply twist its length and your rope is done.

A Note About Hickory: As most young hickories mature, the hard outer-bark develops short fissures that widen as the tree ages, eventually defining recessed diamond shapes with raised cross-hatching bordering the recesses. The raised bark takes on the appearance of a wooden mesh – like a fabric that has been magnified so many times that the pores (diamond shapes) between the fabric threads are visible.

This fabric-feature of hickory's outer bark is a symptom of the fiber-pattern of its inner bark, which produces its sapwood, too. And sapwood becomes heartwood.

Fabric-like fibers = tough wood. This is why so many tool handles are made of hickory.

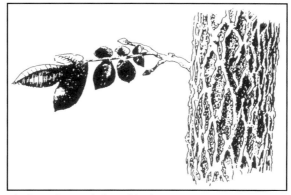

hickory diamond-shaped bark

For many trees whose green inner bark is usable for rope, the fibers run parallel along the branch, and so extraction is necessarily linear, from the base of the branch toward the tip. With hickory it is possible to cut two sinuous lines a quarter-inch apart (never heeding any integrity of grain) then peeling off the inner bark, which is ready, as is, for rope. It's like cutting a strip of fabric along any curved line you choose. Such a random strip of clothing, whether straight or curved, could be twisted and used as cordage.

Leaf Fibers

A dedicated rope-maker looks for fibers everywhere … even in leaves. Substantial cordage can be made from **cattails** (dried then re-hydrated) and grasses (best when green and dampened). But the prime leaf is yucca. In fact, yucca provides one of the strongest natural fibers on the continent. Even dead yellowed leaves found at the base of a living yucca can be too tough to pull off the stem. After removing them with a knife, wet and knead. Twist the leaf and test its strength. Who would have thought that a single dead leaf could not be torn apart?"

Making Yucca Rope

– Taking care to avoid the sharp points on the leaves, cut a green leaf off a yucca plant. Using a smooth dead log as a workbench, gently scrape the outer covering from the leaf with your knife blade held at a perpendicular angle. Then alternately ret (soften by soaking) and gently pound the fibers with a smooth mallet, pulverizing the remaining leaf tissue and leaving behind the yellow-green fibers. Once you have a supply of fibers, arrange them in an overlapping loose string of desired length and thickness. Using a new technique explained below, begin the rope-making process – being careful not to pull the overlapped fibers apart as you hold tension in the two working halves.

NOTE: Yucca leaves contain saponin, a natural soap and mild poison! Though saponin may not pose a serious threat to humans, it is toxic. It's time to substitute a branch nub for the clamp of your teeth. Keep in mind that the nub is now serving as your tooth and you are looking at the process from the other side; therefore, you will twist in the same directions as before but the <u>right</u> hand will cross over this time.

Or simply pinch the loop of the fibers in your left hand (thumb and index) with the loose ends oriented to your right. Grip the far strand with your right thumb and index (holding close to the loop) and twist it *away from you*. Then bring that twisted far strand toward you over the closer strand and re-pinch it in this nearer position. Now twist the "new" far strand and repeat.

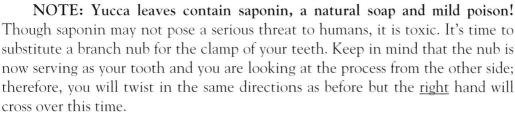

Plant Stems

A good number of herbs (weeds and wildflowers) contain fibers in their stems that are useful in rope-making. Some may need brief wetting, extended soaking, and/or light pounding. Others are not candidates when green but can be used after they stand into winter and dry. These latter need to become almost woody, at which time they can be wetted and gently pounded with a mallet. Sometimes fibers are best obtained by breaking the stem in places and pulling the core of the stem away from the fibers. To get your herbal rope adventures kick-started, try some of these: **dogbane, milkweed, ironweed** (*Vernonia noveboracensis*), **blackberry, velvetleaf** (*Abutilon theofrasti*), and **stinging nettle**. Then experiment on various plant stems as you encounter them and learn the joy of self-discovery. The chances are you could teach yourself a plant cordage that no one before you has written up for the books. Unless you positively identify a plant and research it, **consider your experimental plant toxic (like dogbane and milkweed) and keep it away from your mouth!**

 <u>Rope From Weeds</u> – Depending upon the season, gently break weed stems (green or winter-dried) to see if viable fibers are present for rope-making. If a green plant snaps clean without promise, save part of it and let it season so that you can try again after it has dried. At that point, first try breaking the core away from any fibers. Break it every inch or so. If fibers come off intact, wet them and make rope. If the fibers need softening, beat them gently with a smooth mallet on a bark-less log. If none of that works, wet the dry stem first, then try separating core from fibers and/or using the mallet. Learning rope sources is a study over the seasons.

Vines

Vines, being long, thin, and twining, may appear to be string-like, but most vines are not very useful as cordage, especially when it comes to tying a knot. But there are some vines that can work as cordage for you. **Grapevines** are famously strong and flexible. Very green vine shoots might break too easily, but select, slender air-roots that hang down like harp strings from horizontal sections of stem make wonderful cordage just as they are. In winter I have found them strong and flexible enough to replace a broken shoelace. Mature living vines varying in thickness from 3/16" to 3/8" often demonstrate good strength and flexibility. Obviously, you are not going to accomplish a tight square knot with a thick woody grapevine. You will encounter some weak spots along a vine, but generally speaking – without extremely sharp bending – grapevine will serve you in many a project.

On rare occasions I have found large dead grapevines whose inner bark provided cordage fibers; but, most often, dead grape inner bark proves too brittle.

Perhaps the most impressive vine is **kudzu**, the infamous invasive that was originally brought here from the Orient and planted on railroad embankments to stop erosion. To say the least, kudzu made itself at home in the Southeastern United States, which replicates kudzu's homeland by being at the southeastern corner of a land mass that fronts an ocean. It is a member of the pea family,

compound with three broad leaflets. Kudzu, an uncannily fast-growing vine, is a plague to forestlands of the South. Harvest it freely. (See kudzu photo in the color section.)

Kudzu Inventions

Kudzu Inventions – In summer move carefully through a kudzu patch, an ideal habitat for mice and, therefore, venomous snakes. Challenge students to create an innovative product out of the vine: basket, hammock, awning, table, clothing, swing, etc. Vine strength and flexibility vary with age of the vine and season of the year. Experiment and discover.

SPLICING ROPE

splicing

When substantial ribbons of inner bark fibers are available but not long enough to make a desired length of rope, start the double-helix technique by grasping a cluster of fibers *off-center*. In this way, one of the two twisting strands will run out before the other. Just before this happens, fray the last 2" of the end of this shorter strand so that it looks like a miniature broom. Using your thumbnails make the "broom straws" as thin as possible. Then remove one-half of these "straws" by plucking off every other one at the "broom handle." Choose a new cluster of fibers (same thickness as the original) that will be used to continue the one that has run out. Prepare one of its ends in exactly the same way. With both ends reduced by half their fibers, when you loosely mesh them together, the two halves become a whole again. Handle this mesh site with care as you twist it into one apparent strand. Once embedded in the rope, the splice will not slip … nor will it be a weak spot. You can make an indefinite length of rope by continuous splicing.

Trying Your Hand at Splicing

Trying Your Hand at Splicing – Take two ribbons of equal width from the inner bark of a dead tulip tree. Start making cordage with one ribbon, being sure to start a few inches off-center. When the short end of that ribbon starts to run out, prepare it for splicing and then prepare one end of the second ribbon. Splice and repeat.

Making Clothes

Making Clothes – It's a fact: If you can make rope, you can make clothing. To break into this skill on a beginner's level, weave a *chest-bib*, a sartorial accessory designed to provide increased insulation over the bronchial/lung area in cold weather. All that is needed for the project is lots of tulip tree inner bark and patience. Using the double-helix rope-making technique, make 60 strands of rope, each 18" long and 3/16" thick. Using a flame, clean each rope of its whiskers by singeing. Lay down two strands like a large "plus sign." Using a typical alternating

over-and-under weave, add a strand alongside one of the original two, and then add its perpendicular mate. Continue this until you have 30 strands woven on the x-axis and 30 on the y-axis with loose ends all around. Make one more rope (a strap) to yoke around the neck to hold the bib in place. Tie each end of the yoke to a vertical strand (y axis) at each of two neighboring corners of the bib. All other loose ends should be folded back into the weave of the bib to prevent raveling.

Advanced Weaving
– Make two bibs and join them on three borders by tying together the loose ends of one to the other. Into the open side stuff dead grasses, leaves, cattail down, or kneaded fluffs of dead inner bark. Then tie off the loose ends of that fourth side. Now you have a warmer bib, a pillow, or a simple pouch full of tinder.

Primitive Needle and Thread
– Without altering the sharp end (2 ½") of a green yucca leaf, wet and mallet the rest of the leaf into a mass of fibers that remain connected to the tip. The fluted end and spine will serve as a needle

to guide the twisted fibers (thread) through a pre-punched hole. This combo provides a quick solution for clothing or gear repairs.

Animal Cordage

Cordage can be made from animal skin, sinew, fur, or gut. Rendering animal skins into rawhide and buckskin is covered in *Secrets of The Forest, Volume 3*. Here are the fundamentals of three cordages made from animal parts.

Making a Rope from Rawhide
– Trim a stiff hide to the shape of a rough oval or circle, and then start anywhere on the edge to cut a 1/8"-wide strip that spirals toward the center. Soak the strip in water until it is flaccid. Tie one end to a high tree limb and the other end to a stick and, keeping the line taut, rotate the stick propeller-style to turn the wet strip into a tight twist. Anchor the stick on the ground with a rock to keep the line from kinking or unwinding. Let dry. When later you make use of this rope, you may want to soak portions where knots will be tied in order to achieve tight turns in the rope. The rawhide will shrink as it dries.

Making Buckskin Rope
– For a softer version of the same rope, take the skin through the complete tanning process and then cut a spiral strip from the finished buckskin. Either type of skin rope is strong enough to be used as a bowstring.

Making Sinew Rope
– Sinew is a silver-white sheath of connective tissue that surrounds certain muscles and tendons in mammals. Large pieces of sinew are the only sources worth working, so large animals necessarily supply this resource.

In the modern-day East, where the buffalo do not roam, deer (but not antelope) still do play and supply us with sufficiently large sheets of sinew along the Achilles tendon and over the spine from shoulder blade to lower rib cage. Expose these enveloping sheaths by prying them out with the dull side of a knife blade. Sever the longest pieces and scrape away their surface material with a sharp knife, blade held perpendicular to the sinew. Allow to dry in sunlight until the tissue turns beige and as rigid as hard plastic. Gently pound it with a rounded river stone on a smooth, bark-less log until the beige material breaks away leaving bright white clumps of thread.

These threads are very strong and make their own glue. Saliva softens the threads and activates the glue. The limp wet threads can be wrapped onto surfaces like arrow shafts for attachment of fore-shafts, nocks, points, and feathers. No knots are needed. Simply twine and press the sinew upon itself. As it dries the glue hardens and fixes the sinew cordage in place. However if it gets wet again it will loosen. Sinew's historical uses are many, but perhaps none so famous as its application to bows – both as a bowstring and in "backing" a bow to strengthen the wood and to prevent splintering.

"There is nothing so soporific as the sound of rain at night ... unless it is dripping on your forehead." ~ *a first-time survival student on the day after*

CHAPTER 14
Oh, Give Me a Home ...
~ Shelter Building ~

SHELTER

There are three thieves that rob our bodies of heat: air, water, and earth. If air is moving, it carries heat away faster than stagnant air. Water carries away heat many times faster – especially if there is current ... and, of course, the rate that heat is stolen from the body increases as the air or water temperature decreases. The fact that the ground beneath our feet absorbs heat from us becomes critical when we expose more of our bodies' surface area to the earth, which is why lying down on bare winter ground makes us cold.

Once a body's core temperature drops below a crucial threshold, sustaining body heat becomes impossible without the outside help of a fire, warm liquids ingested, or contact with another warm body. While an open fire gives us the most immediate relief from the cold, rain and wind can compete with your comfort level as you sit around that fire. Shelter provides a reliable, long-term solution to retaining body heat, because it can block out wind and rain. In a sustained driving rain in cold weather – though you can successfully keep an outside fire going by covering it with a tightly packed "tipi" of wood – such a pyre covered by the outer wet wood might not throw off enough heat to override the chilling effects of the weather. Even better than shelter or fire is a combination of the two, but great care must be taken to prevent the fire from engulfing the shelter. A primitive shelter made of dead wood and dead leaves might be described as "a giant pyre" ready to be lighted.

We shall begin with a winter shelter that does not utilize a fire – the kind of shelter to be built when the survivalist is blessed with dry weather but is not confident about making a successful fire. (For one who is about to construct a survival shelter, a wet forest *demands* a fire, which demands a shelter that can protect the fire. Wet leaves should never be used as insulation in contact with the body.)

Native Americans of the Eastern U.S. watched gray squirrels build their nests of sticks and leaves, and from that observation they borrowed the squirrel's ideas and put them to their own use, modifying the architecture to better accommodate the human body. A squirrel makes an orb-shaped home, which requires sleeping in a balled-up position – like a dog curled up, nose to tail. Humans are not apt to remain in such a position all night.

The gray squirrel naturally uses the materials that are most abundant. It takes many a trip up and down a chosen tree to carry all that material, but the squirrel's work is easier than that of a human, who is undertaking a larger construction on the ground. Such a survival shelter must provide an extraordinary amount of insulation: from the ground, from wind, and from rain. If we had the perfect coat of a squirrel and that luxurious tail to wrap around us, and the flexibility to sleep all night in the fetal position, our workload would be lessened. Since we are not so blessed, we have to compensate with a "truckload" of leaves. Thankfully, unlike the squirrel, we do not have to haul materials in our mouths to our work site.

First Tool

An energetic, dedicated survival student can in four hours build a survival shelter designed to envelop and protect the body from rain and winter's cold. Most of this time is spent gathering leaves. Therefore, a worthwhile investment of your day is to make a rake. It takes less than thirty minutes to make a rake that can cut the leaf-gathering time in half.

Making a Rake
– After each of your students has made 2 strands of 2'-long rope, divide your class into groups of three or four and give each group the challenge of making a usable rake. Give no instructions. Let the design and manufacture arise completely from their ingenuity. Even though modern humans are savvy about rake-design, the making of the tool is a unique experience. This project is, in all probability, the first composite tool they have made of all-natural materials. The accountability is wonderful. Either the rake works well … or it doesn't.

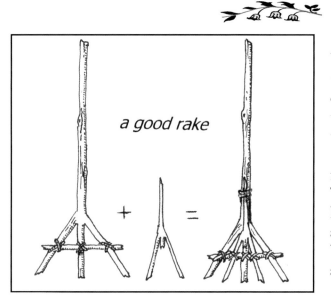

a good rake

While the rake-making is going on, put together your own original rake or one based upon the diagram to the left. Notch the parts that cross-contact (tines to brace) and where rope will lash parallel pieces. This notching and lashing will keep those parts from shifting and sliding upon one another. When finished, set your rake aside out of sight.

<u>**Rake Race**</u> – When all rakes have been completed, let each team of designers demonstrate the efficiency of its rake. Let all students critique each design. But the ultimate assessment of a tool must be its efficacy. Hold a contest on raking up enough leaves to fill same-size trash bags.

<u>**Making "The" Rake**</u> – After the above contest, make a group decision about which design proved most efficient and sturdiest. Now repeat the rake-making challenge, this time with all groups mass-producing the same kind of rake. If no prototype proved inspiring in the competition, show the teams your rake (and the design illustrated above) and have them replicate it.

The value of a rake increases when there is light snow on the ground. (In heavy snowfall, snow would be the shelter-material of choice.) Leaves beneath snow are prime shelter material if they are still dry. Getting to those leaves under the blanket of white becomes a mission with an important consideration: keeping yourself warm and dry. A rake allows you to do this, because not only does it scrape away snow and gather leaves … it also serves as a leaf carrier. Cold, wet hands (or clothes) can debilitate and lead to hypothermia.

Shelter Site

Your site selection should take into consideration the availability of bulk insulation material. If broad-bladed leaves are not available, you'll have to find a substitute. Perhaps grasses are available. If so, choose a site near the tallest grasses. Or if evergreen needles are to be your main supply, set up your shelter close to a source where branches can be reached.

Other site selection factors include:

1. Water runoff – Don't build where water will channel or pool during a rain.

2. Wind – Stay below a ridge top. Make use of natural barriers against prevailing winds.

3. Suspended humidity – Don't build in the bottom of a valley where cool, moist air pools on a nightly basis.

4. Future deadfalls – Check for "widow-makers" (dead trees or limbs that might fall).

5. Animal homes – Don't invade an obvious territory, especially where a burrow is evident. If a strong musky scent of urine is present, choose another site.

6. Sun versus shade – In winter the south side of a mountain is best, then east, then west, but never north.

<u>**A Place to Call Home**</u> – Walk the forest, taking your time to find that "right" place for your shelter. Assign to different students specific dangers or concerns that could negate the feasibility for a site. Have each investigator report his/her findings for each proposed site.

The Ridgepole

Find a dead tree trunk to serve as the backbone of your structure. It must be strong enough to support your shelter when the leaves are wet (say, 200 lbs.), long enough to slope to the ground and cover your prone body and more. If a good candidate is too heavy to carry, consider building your shelter where that log lies. For young students, transporting a log to a chosen site might be the great adventure of the day.

 The Spine of the Shelter – After deciding upon your site, locate a fallen tree (or standing dead tree) whose trunk will provide a suitable ridgepole – say, 14'-20' long. Break off limbs by hand and then finish off limb-nubs with a weighty rock-hammer. Test the log's strength by propping it up at an angle and applying 200 pounds of body weight(s) to the center point.

The Ridgepole Support

It is helpful to find, somewhere in your chosen area, a natural convenient support upon which to rest one end of the slanting ridgepole of your shelter. Consider a thick fallen tree trunk, a boulder, or a low fork in a standing tree. If such a support is not available, break to size a strong forked branch as a prop. Settle one end of the pole into the fork and lean the prop and pole against a standing tree. Otherwise, a free-standing tripod prop will be necessary. To create such a tripod of stout sticks, each stick must have a fork at one end.

 Making a Tripod Support – If your students can make a successful tripod that can support all the weight of a shelter, they can feel confident about building a shelter anywhere. Divide them into groups and give each the challenge of making a tripod design with 3 found stout sticks. (If a group cannot envision a mechanical solution, help them get started by offering 3 plastic picnic forks. See how long it takes for someone to realize that the potential for a tripod is best with the fork tines up where they can mesh, not down on the earth.) Each stick should have its supporting crotch the same height off the ground – say, 3 ½ feet. "Trial and error" will surely be the name of the game. Budding architects and engineers might emerge from this exercise.

 Forest Gymnastics – For tripod-accountability instigate a game of "slanted balance beam." Who can walk up the newly supported ridgepole? For safety, have spotters walk along each side. Remove any objects on the ground that could injure. (Spotters should not attempt to catch a falling gymnast. Instead, they should offer horizontally raised forearms to be grasped like a banister railing merely to help the gymnast break his fall.)

More Forest Gymnastics

More Forest Gymnastics – For more daring fun in tripod product-testing, have each group build close enough to the others so that every possible pair of tripods can be spanned by a ridgepole (this time horizontally, not slanted). Take turns letting students walk the pole from one tripod to the other. Again, for safety, set up spotters on either side.

Make sure that both the support and ridgepole are strong. If the best available ridgepole seems too long and is unbreakable, consider using only a fraction of its lower end for the shelter. The excess can jut out unused. (Later, you may use it to build an awning over a fire area.)

The angle of the ridgepole to the earth is determined by lying down on your side and positioning the ridgepole 6"-8" above your shoulder and the same above your feet. (The heavier the shelter-occupant, the less clearance measurement … due to increased compression of leaves beneath.)

Setting the Ridgepole

Setting the Ridgepole – Determine the pole's proper angle by lying beneath it and varying its length by positioning the ground-end farther from or closer to the support. The orientation of the shelter is a consideration. If you can determine the direction of prevailing winds, situate the low-end toward the wind. If the wind is variable, position your shelter so that one of its sides will receive a lot of sun.

Walls – the Infra-structure

The first **A-frame sticks** will not only support **shingle-leaves** on the outside of the shelter cavity but also trap **bedding-leaves** on the inside. Gather lots of different lengths of straight dead sticks – preferably with some branching. You'll want to completely break off all secondary branches except along one facet (rank) of each stick. (Any sub-branches pointing toward the cavity could guide water droplets into the shelter.) The 1"- to 2"-long branch nubs you leave on the outside will serve as brackets upon which to build a **cross-lattice** of thinner sticks.

Building Walls

Building Walls – Stack an A-frame of partially trimmed sticks along either side of the ridge-pole at an angle of 60° – 65° to the ground. Position the branch nubs outward. It is best not to let sticks cross at the top (above the ridgepole) like tipi poles, because such a cradle will collect water and channel it toward the ridge-pole and, thus, the occupant. Don't squander the long sticks you find by breaking them for the low end of the shelter. Use long sticks only for the tall end, because long straight sticks are not as plentiful as short ones. Make each wall a perfect plane. For a solo shelter make the A-frame base about 3'4" wide at the entrance (high end of the ridge pole). This dimension varies, of course, by body size. For a shared tandem shelter, make the base wider and the ridgepole higher. Due to

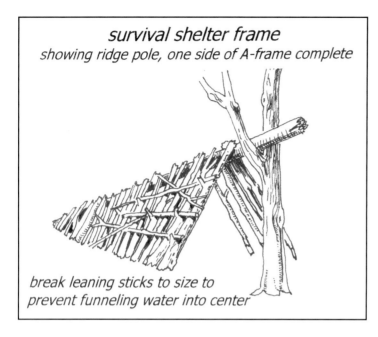

survival shelter frame
showing ridge pole, one side of A-frame complete

break leaning sticks to size to
prevent funneling water into center

shared warmth, the tandem shelter is the preferred scenario in cold weather. Lay down enough A-frame sticks to mostly conceal the cavity. When you finish, double-check that no branch nubs are pointing toward the cavity.

The Lattice

To complete your walls, add long, thin sticks and vines across the A-frame nubs at horizontal and diagonal angles. Not only do these crosspieces cover gaps in your walls but they also provide more traction for the first-stacked layer of shingle-leaves.

__Making a Lattice__ – Lay sticks horizontally and diagonally on the A-frame nubs criss-crossing the walls to fill all gaps. The finished A-frame defines the sleeping cavity, supports shingle-leaves that cover it, and traps bedding-leaves that fill it.

__Millions of Leaves__ – Use the rakes to pile up leaves where they have accumulated in deep drifts. Use arms, clothing, tarps, rakes, or anything else you can find for hauling leaves to the shelter site. Do not rake the shelter's uphill side-yard. That ground-cover protects the abode from rain run-off. Save the driest leaves in a special pile as bedding-leaves for the interior.

Shingles

The single-most important waterproofing factor lies in how you lay down the outside leaves of the shelter. Leafy units called "shingles" must be stacked in imbricate fashion at an angle designed to drain water away from the shelter's interior. Slide your forearm (palm up) under a very large mound of leaves (at least 2' high) to take an armload. With that forearm flat on the ground, tamp down the top of the orb of leaves with other hand and forearm, flattening leaves into a single "shingle." As you compress the leaves, vary the orientation of the tamping arm like a second hand on a clock and tamp at least 12 times. Bulk-wise it's like changing a glob of dough into a biscuit shape. This simple manipulation takes only a few seconds per shingle but goes a long way in *aligning the majority of leaves* into a flat unit that will effectively work like a shingle. This is the key technique for keeping your shelter dry.

Lean the shingle, half on the ground and half on the bottom of the A-frame, so that it and all subsequent shingles will tilt away from the hut. The desired an-

gle of tilt lies between 30° and 40°. Once shingles have been laid down all the way around (except across the entrance at the high end of the A-frame), begin a second tier that leans half on the top of the first row and half on the A-frame above the first row, maintaining the same tilt-angle. Once you have stacked the first wall all the way to the top, begin a second wall of shingles that will cover the first, again being sure to maintain the proper tilt-angle. Add more walls until the shelter shows a thickness of 2' all the way around. Many of my students convince themselves to settle for a dark cavity, as if their goal is to exclude light rather than rain. With the experience of a rainy night as a teacher, they spend the following morning adding more leaves to achieve the suggested 2'-loft.

Laying Down Shingles – Haul leaves to make two big supply piles – one on

either side of the shelter. Lay down the first tier as a base and then build upward. With a class of students working on one shelter, this job can done in less than an hour. Impress upon the workers that the walls need to be **at least 2 feet thick.**

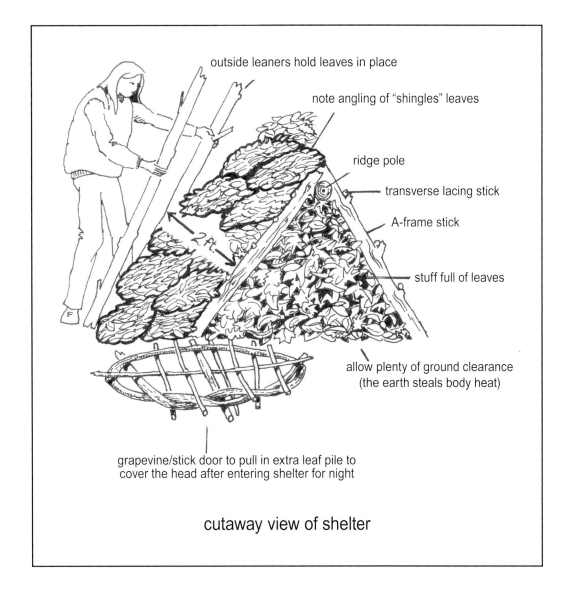

outside leaners hold leaves in place

note angling of "shingles" leaves

ridge pole

transverse lacing stick

A-frame stick

stuff full of leaves

2 ft

allow plenty of ground clearance
(the earth steals body heat)

grapevine/stick door to pull in extra leaf pile to
cover the head after entering shelter for night

cutaway view of shelter

More Leaves

Two more piles of leaves should be held in reserve to complete the shelter. One pile (your driest leaves for bedding) will fill the interior. The other will be used to "close the door."

Outer-structure

By leaning hefty sticks against the newly made leaf-walls, make a second A-frame outside the shingles to secure them against strong winds. These sticks compress the leaves somewhat, making smaller air pockets (and, therefore, a warmer home). These air pockets will serve as secondary insulation for body heat retention. Primary insulation will come from bedding-leaves. Again, don't cross these outer A-frame sticks at the top. Break them to the correct length to avoid water build-up at the ridge pole. These sticks can be thick (all the better to shed rain), and their weight will provide substantial strength to the shelter for that crucial moment when you auger your body into the bedding leaves.

interior of a shelter

 <u>Securing the Leaf Walls</u> – Lean the outer A-frame of sticks against the leaf walls to hold the shingles in place. Use all available long sticks on the higher end first!

Bedding-Leaves Make a Sleeping Bag

Cram the cavity with dry leaves, this time taking pains to remove sticks or cones that might prove uncomfortable during the night. Stuff as many leaves as humanly possible into the space. This pile of leaves will be your primary insulator – mattress and blanket – providing thousands of air spaces that trap body-heat. Later, when you're ready to retire, auger feet-first into this pile, being sure to slip in at mid-height to ensure adequate leaf-loft below to insulate the body from the ground and above to slow down (and back-up) the rise of escaping body heat. One of the most common mistakes made by shelter-users is slipping into the top of the cavity, thereby having no bedding-leaves above the body for insulating all that upward escaping body heat.

 <u>Completing the Shelter</u> – Stuff the shelter cavity with as many dry leaves as possible. You can't overfill! On the contrary, in the middle of the night, when your interior leaves have compressed from your body weight, you may wish you had been more industrious stuffing leaves for your bed. Prepare yet another leaf pile outside for closing the door.

The Door

A last pile of leaves left by the entrance should be large enough to cover the entrance. This pile will serve as your door. Since more than 75% of your body heat is lost through the neck and head, this collection of leaves is critical. This door-pile should be close enough to the shelter so as to reach *through* it with an arm.

To simply try to pile leaves on your head is a futile exercise. No matter how well you manage it, you are certain to wake up in the night, your head exposed, and the leaf pile scattered. These "door-leaves" need a "door-lock" to hold them in place.

Making a Grapevine Door-Lock

Making a Grapevine Door-Lock – Cut a 12-foot section of green grapevine ½" to ¾" thick. Bend it into an oval 4-feet long and 3-feet wide, twisting the loose ends around the oval so that they lock in place by their own pressure. Such an oval should be approximately large enough to cover the triangular entrance of the shelter, though it will never actually contact the entrance.

grape vine

Lay this oval flat on the ground as you kneel before it. Position 3 or 4 straight branches along its long side like bars on a window – the first bar on top of the oval, the second underneath and so on, alternating each time. Now using horizontal bars, lock the vertical bars in place by weaving sticks over and under, providing their own pressure to hold the vertical bars against the frame. These horizontal bars need to be fairly flexible to accommodate the weaving. Green branches work best, but dead branches can serve if they do not break in the bending required for weaving. Sections of thicker grapevine also work well for these bars. Start your first crosspiece under the left side of the oval frame and have it run over the first vertical bar (which was laid on top of the oval). Then work the crosspiece under the next bar, which was laid under the oval. Or, allow a looser weave as seen in the illustration. Add enough crosspieces to make a sturdy "door lock."

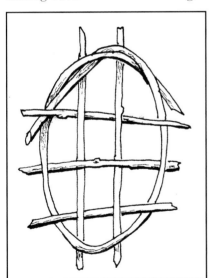

grape vine door

Locking the Door

Locking the Door – Pile the door-leaves close to the entrance with the grapevine door-lock leaning against them on the opposite side from the shelter. As you auger into the bedding-leaves, reach through the door-leaves, grasp the door-lock, and pull the pile of leaves toward you to seal the entrance. The weight of the leaning lock will hold the seal.

Time-Share (more leaves)

When partners time-share overnight in a shelter, each arriving sleeper must have prepared a ready pile of bedding-leaves to re-stuff the cavity. The previously -used interior leaves will be compressed from the last tenant's body weight. The pile of door-leaves will undoubtedly be scattered from the last exit, so care must be taken to pile them up again.

A Taste of Winter – On a camping trip in cold weather, have your students set up tents and sleeping bags as usual with the understanding that the mission of the trip is to experiment with primitive shelters. Ask how many students would like to spend a part of a night in a shelter. Once you have that number, you'll know how to divide your class into shelter teams. Four students can share one shelter on 2-to-3-hour intervals. (If only 1 to 4 students accept the challenge, then only one shelter need be made.) Each intrepid shelter-sleeper should bring a watch with an alarm and set the alarm for the chosen time. When the alarm goes off, that sleeper must come out and wake up the next shelter-sleeper. This new tenant then walks by flashlight to the shelter, re-stuffs the cavity, fluffs up the door pile, enters the shelter, reaches through the fluffed door pile, and pulls the lock to lean upon the door. In 2-3 hours, when the alarm sounds, that person exits the shelter and wakes up the next person. And so it goes.

The Downside of Nocturnal Urination

Getting in and out of the shelter multiple times in one night is counterproductive. You'll likely never arrange that pile of door-leaves to cover you as well as you did the first time you entered. The bedding-leaves will not enfold you as well as they originally did. This makes nighttime urinating a problem. Ideally, urinating just prior to retiring can preclude the need to get out before morning, but in reality the need to exit is common, especially among old codgers.

There is a solution for males. (I leave it to an enterprising female to invent a method for ladies.) Stripping outer and inner bark off of some dead trees (like the tulip tree) produces long strips shaped like "half-pipes." A section of this bark laid on the ground and running through the wall on the downhill side of a shelter can serve as plumbing – a conduit for urine. Only accuracy is required.

Shelter-Pirates

A well-made A-frame shelter can stand for years, but unless you are actually using it on a daily basis, **never enter an abandoned shelter**. When a creature of the wild encounters an abandoned shelter, it must be difficult to pass up such a gift. In taking apart the shelters of students weeks or months after they had used them, I have encountered the following: venomous snakes (twice) skunks (thrice), and, cicada killer wasps (once).

A Hearth in a Shelter

If a natural rain shelter can be found, by all means use it. Such a discovery can save a lot of work. An overhanging rock ledge is a treasure. So is a cave, though humidity and mold is problematic. (I lived in a cave for a time, but had to abandon

the abode due to bronchial inflammation.) Each of these shelters can safely accommodate a fire. If such quarters cannot be found, and if an inside fire is desired, a shelter must be built on a grander scale, simply to prevent flames from igniting shelter materials.

There are two scenarios that make imperative the building of a shelter large enough to utilize a fire for heating: 1.) When all forest materials are wet, and 2.) when biting insects are in season, yet the nights are cool. In neither case do you want to immerse yourself into leaves.

For safety, the size of a shelter designed for a hearth must be much larger than the A-frame previously described. The fire will provide any needed warmth, so less work goes into insulating from the cold and more time goes into rain/wind-proofing. A much larger version of the A-frame can be used without the interior stuffing. Or a simple, single-plane, lean-to roof can be constructed with walls positioned on the two sides to shut out the wind. Build a wind-break on each side by driving two rows of "stockade" sticks into the ground, making a double-fence about 1' apart. Fill each double-fence with dead leaves.

The high end of the shelter, which covers the fire, should be at least 6' above the ground. This hearth-side is left open; that is, there is no front wall. During construction it is difficult to reach so high for the placement of shingles of leaves, but leaves are not necessarily needed there. Slabs of bark can cover the eave over the hearth. Or the hearth itself can include a capstone for protection. Position a big stone slab or a wall of smaller stones at the backside of the hearth to reflect heat into the shelter.

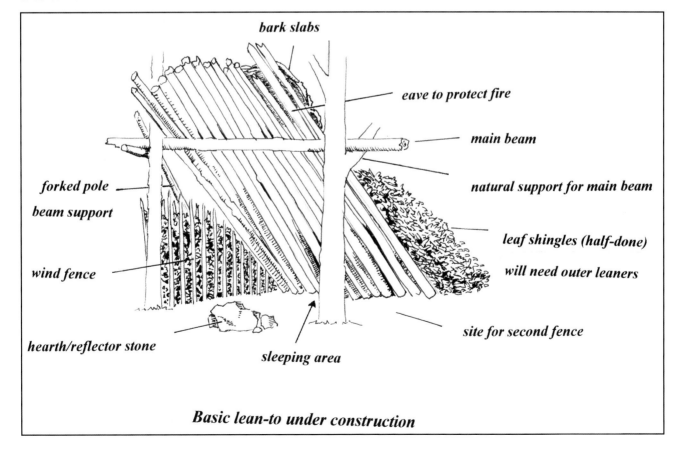

Basic lean-to under construction

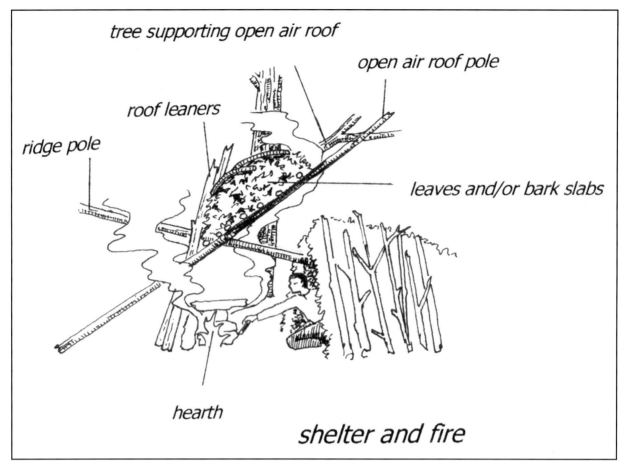

tree supporting open air roof

open air roof pole

roof leaners

ridge pole

leaves and/or bark slabs

hearth

shelter and fire

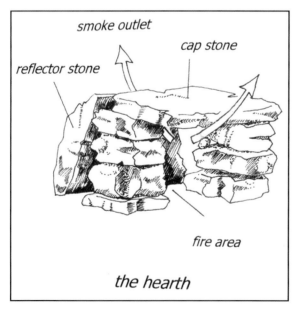

smoke outlet

reflector stone

cap stone

fire area

the hearth

Another type of fire-cover – the open-air roof or eave – can be constructed over the hearth as shown in the diagram. Fuel can be stored inside the hut.

Inside shelters that utilize hearths, leaves are needed in the sitting and sleeping areas to insulate your body from the earth. A rectangular "bed-frame" of logs can keep these leaves from scattering. In summer in the South, because of the presence of chiggers, the choice of your bedding material is critical. Consider the following: wood shavings carved from green sassafras and juniper, green leaves of pawpaw, black walnut, perilla, hickory, hazelnut, juniper, sassafras, and bracken fern – all containing insecticidal properties that discourage invasion from below.

 Build a Rain-proof Lean-to – Using the illustration as a guide, build a single-plane lean-to with side walls of double-fences. After completion, keep

an eye on the weather and plan to be in your new home when rain arrives. It's a bonding experience for a group of students to build a lean-to together and then sit under its protection when the rain comes down. As always, the great teacher of accountability will be present in the scenario.

Arched Roof Shelter

A roomier enclosed shelter can be made with arched ribs for a frame. You'll need plenty of cordage for this. Because the peak of the roof is rounded, the most difficult aspect of this design is rain-proofing. Once, inspired by the discovery of a large tulip tree that had recently fallen, I built such a shelter using large slabs of bark as shingles. (A frame must be strong to support bark when it soaks up rain.) Found litter becomes a treasure: an abandoned raft or tarp or roofing-tin blown to the hinterlands by a tornado. When useful litter enters your life, use it and be grateful.

<u>Arch-Frame Construction</u> – Hammer a 2'-long, 2"-thick hardwood stake
8" into the ground, gently work it free (so as not to overly widen the hole), and pull it out. Do this sixteen times in two rows of eight, 6' between rows and 1' between holes. Make these two rows of holes slightly angle away from each other to increase the curvature of the arches-to-come. Cut sixteen maple saplings, trim them, and ram their thick ends down into the angled holes, each stave standing (tilted) about 8' above the ground after being inserted. By bending opposite poles (from opposite rows) toward one another, lash together eight arches and then a light ridgepole, like the ribs and keel of an inverted canoe.

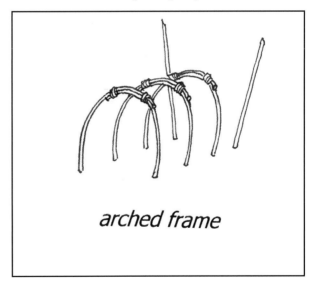

arched frame

From each end of the ridge-pole wedge a long forked pole diagonally to the ground to prevent collapse of the shelter to the front or back. Tie cross lattices lengthwise to the outer sides of the inverted "canoe" and across its "stern." The spacing between lattices depends upon the material you will use for the walls. (For piles of leaves consider 6" intervals; for thatched grass, base it upon the length of the grass. Then begin a marathon roundup of leaves or thatching to build up the sides and back. Lean sticks against the leaves (or impale the sticks into the ground) to hold them in place all the way up to the curve of the arches. For the rounded roof use: 1.) large slabs of bark stacked in an imbricate fashion, 2.) bundles of grass tied and thatched, 3.) compressed boughs of evergreens oriented with all their needle-clusters directing water from the shelter, or 4.) any found trash that will shed water.

Near the shelter entrance build a stone hearth with an ample slab of rock propped over the flame. If necessary open a ventilation hole in the top of the back wall of the lodge for a smoke draft. If you've properly gauged the prevailing winds, most of the smoke will escape out the front entrance and the hearth will throw heat back into the shelter.

Shelters with Hearths

A large lean-to shelter with an eave over the hearth in front poses a construction problem: The eave must be necessarily high for fire-safety. That height makes it difficult to lay down roofing material. One solution is to make a scaffold.

 Building a Scaffold – Few exercises of teamwork can top the project of constructing a forest abode and then using it. Begin the hearth-shelter project by constructing a scaffold. Rope, bamboo or tree branches, and ingenuity are the raw materials needed for this project. Challenge a group of students to reach (by hand) a bag of fruit suspended from a limb 12 feet off the ground. When someone reaches the treasure, the prize is doled out to all.

Cone Shelter

Another design for a roomy shelter is a wooden tipi. Begin with an 8'- high tripod, and then simply lean long sticks into a cone shape. Stack the poles as tightly as possible all the way around except for the allowance of an entrance. Add to the exterior of the tipi a "coating" of tied bundles of grasses or clusters of ever-green boughs shingled from ground-level to the apex.

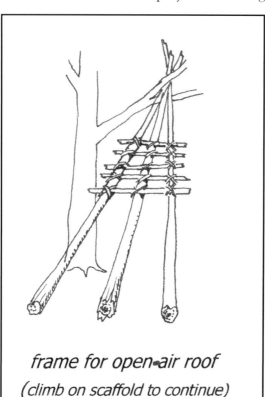

frame for open-air roof
(climb on scaffold to continue)

I built such a shelter in Washington State using lots of compressed green boughs of northern cedar (technically, a cypress) for shingling. It proved 100% rainproof after a storm. This project required no cordage! All greenery was hooked over latticework that was suspended by limb nubs. The resulting abode looked like a living organism, lush and verdant and camouflaged.

 A Verdant Tipi – Construct the tipi frame with long, straight branches or logs, except at the entrance, where crooked tipi-poles can afford a doorway by

making use of a dog-leg bend in a few sticks. Leave limb nubs on the tipi's exterior. Cover the cone with a lattice of sticks and vines. Depending upon what materials are most available for the tipi's "cover," gather lots of green shingling that shows a uni-directional orientation of its needles or blades. (Using grass requires making lots of cordage for bundling sheaves. Evergreen boughs can be draped over the lattice.) Once you have covered the tipi with substantial covering, additional bough-shingles can be simply inserted into place at an angle. Dependable scaffolding may be part of the agenda to cover the apex. This shelter is only feasible where supplies of tall grasses or evergreen boughs or bark slabs are abundant.

> *Warning:* **There is an obvious potential for disaster when maintaining a fire inside a shelter composed of flammable materials. Great care must be taken to keep flames well away from leaves and sticks. If you have to leave your shelter for a time with a flame or coals still alive in the hearth, you must be an experienced hearth handler with the knowledge of how to prevent a flame from growing in your absence. Until you become adept at** *banking* **a fire, reduce the fire to coals before exiting.** Fire-banking is covered in *Secrets of The Forest, Volume 2.*

FORT APACHE: A Quicker Shelter for a Late Start

What about a situation in which nightfall is coming on fast? I've heard people talk about crawling into a pile of leaves should an emergency arise, but if they ever tried this they would be disappointed. It takes an enormous pile of leaves for success with a haphazard pile of leaves. You've got to have lots of leaves beneath you to insulate from the ground, and you'll need so many leaves above you to trap rising body heat that the necessary conical mountain of tree litter would stand at least 7' high and spread to a diameter of about 12' at ground level. Furthermore, unless you have trained yourself to sleep without body movement (a "behind-enemy-lines" discipline), the leaves above you will slowly trickle away, exposing you to the cold night air.

Changing that loose pile from a cone-shape to a near-cylindrical shape (by using retaining walls) drastically reduces the number of leaves needed and maintains pile structure. Such a shelter design can be achieved within an hour and a half by an enthusiastic worker. Like the basic pile-of-leaves technique, this design does not protect from rain. For that reason it would be best to build it underneath a natural shelter such as a thickly-leaved evergreen (preferably a hemlock, spruce, or fir in the mountains) or the leaning trunk of a grandfather oak or other hefty tree.

Or perhaps good fortune provides you with some kind of discarded man-made material that repels water. Even in fairly remote areas I have found: a sheet of polyethylene plastic, deflated raft/inner-tube, deflated sleeping mattress, lumber, plastic garbage bags, abandoned tents, and metal roofing carried away by a tornado.

On several ramblings I have found abandoned canoes and john-boats, sometimes damaged, sometimes not, always useful in repelling water. (Rivers are often treasure troves of unexpected resources.) Such finds can be simply laid on top of the finished structure (at a runoff angle) with sticks weighting them down, if necessary. On one occasion, after finding a dead tree from which I could disengage huge slabs of bark, I built a slanted, latticed shelf of sticks across two strong limbs of a living hemlock tree. Onto this shelf I stacked an imbricate roof of bark shingles that did a pretty good job of shedding water. The bark eventually absorbed so much water I was forced to shore up the shelf with forked-pole supports.

"Fort Apache," as I named this abbreviated abode, is little more than a glorified leaf pile, but its all-important improved feature is the inability of the leaf pile to lose its shape. It begins with a roundish or oval palisade of sticks (each about 1"-thick) impaled into the ground to form a leaf-retaining perimeter with a diameter of 6'. Except at a 2'-wide step-over doorway (actually, a 2'-high threshold) each stick of the fortress should stand 5'-high. All palisade sticks should lean slightly inward toward the center of the sleeping area.

As you gather and size dead sticks for the outer-structure, you'll appreciate those limbs that break with a naturally pointed end. If they do not, you'll need to sharpen one end so that each stick can be impaled 6" deep into the earth. (Refer back to the Arched-Frame for hole-making.).

Palisade sticks should stand 2"-3" apart. If time permits, the finished unit can be strengthened by a rough "belt" or interweaving of grapevines several times around the oval. If you can harvest an abundance of vines, fewer palisade sticks can be used, spacing them wider, and filling the gaps with more rings of vine. This offers a faster version of the shelter. Sections of grapevines or sticks can also be stored near the door and used to close off the entrance after you have entered for the night.

With the shell of the fortress complete, lay down a bedding of fresh evergreen boughs inside, one layer running in one direction, the next layer running perpendicular, and so on … until this foundation provides a level loft 12" off the ground. This 1'-clearance, though it will compress from your body-weight, will keep you away from the warmth-thieving earth.

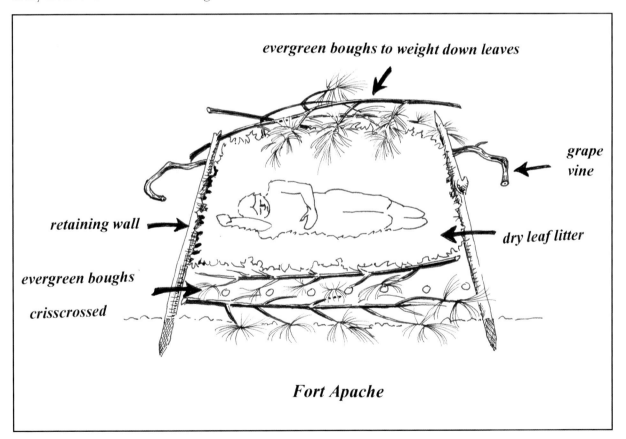

Fort Apache

Into this fortress dump armloads and armloads of dry leaf litter until the structure is brimming over with leaves. Then go inside and lie across the leaves at varying angles to compress them. After this compression, complete the leaf-filling again. (If you are blessed with found water-proofing material, add that now on top of the leaves *at an angle* to allow for runoff.) To keep the leaves from scattering from wind or from your restlessness, lay down a final layer of evergreen boughs. (Dead sticks can be used if they are multi-branched, so as not to roll off to one side.)

Now you are ready to enter. Step through the entrance and stand just inside the threshold. Close off the entrance with branched, horizontal limbs laid one on top of the last. Then bore into the leaf pile at the midpoint, allowing 2' of leaves below (which will compress according to your body weight) and 2' above. You'll want to stay somewhat curled up on your side throughout the night.

One of the advantages of this bedding is that you can urinate without leaving your sleep site. Simply use your hand to auger a tunnel straight down to the evergreen boughs. When you finish, close the tunnel with leaves.

Insulation and Other Warm Thoughts

Besides drawing warmth from the radiant heat of a fire, a chilled person can benefit from drinking heated water. Still another way of accruing warmth is to receive it from another warm body. In a survival situation, two people can take advantage of this … especially in their sleeping positions. If another body is not available, fire-warmed stones make good companions.

Insulation can boost the efficiency of your clothes by surrounding your body with dead air pockets that improve the heat-retention quality of your clothing. Cattail down can be stuffed inside a shirt, pants, socks, shoes or hat. The "hot-dog-on-a-stick" female flower head (once it has gone to seed) is a surprisingly copious source of compact soft, silky fibers. Break off a stalk and, after grasping the seed head in both hands, twist, and watch the downy pappus appear to "foam" out. As you wear the shirt, stuff the fluff into sleeves and the torso area, simulating a fiber-filled jacket. Because cattails multiply by asexual means, they are found in colonies and can usually provide all the insulation you'll need. **Keep in mind: pappus is flammable!**

Pearls on Feathers – To see the remarkable insulating quality of cattail down, twist a seed head of fluff and make a nest of the down in your hands. Have someone drip water into the nest and watch the droplets bead up without soaking in. When this down is "worn" against the body, body heat drives moisture outward through the fluff that will not absorb it. This removes the moisture away from the skin so that it does not steal body heat during protracted evaporation.

The Weather Bird

At the onset of building a survival shelter, one must decide which style of shelter is best. Besides the choice of building materials, a main consideration is weather. If it is raining or if rain seems imminent, it would be best to build a shelter that can accommodate a fire. If, however, the rain might soon let up and the sky clear enough to dry the top layer of leaves on the forest floor, one could construct a smaller hut in which to nestle into those drier leaves (added last to the interior).

As unlikely as this may seem, I have learned to trust the word of one bird for a forecast. The Carolina wren's weather prophecies have for many years proved

uncannily reliable. I don't mean simply hearing the wren's cheerful "galloping" call (variously described as "Canada, Canada, Canada" … or "tea-kettle, tea-kettle, tea-kettle" … or "liberty, liberty, liberty" … or "cheery, cheery, cheery") somewhere out in the forest. For the birdsong to be taken as a forecast, it must be delivered at close quarters and directed at you, as if the wren has singled you out for an ineffable need to communicate.

This friendly burst of notes first got my attention during a rainstorm. With rain steadily drumming on leaves as the only sound in the forest, suddenly the wren made its call very close to me. It was such a startling juxtaposition of sounds – droning rain as background to the melody of a songbird – that I stopped what I was doing and looked for it. By the immediacy and clarity of the notes, I knew that the bird was facing me. When it called again, I spotted it only a few yards away, perched on a branch, and looking right at me. Within fifteen minutes the rain stopped. Within another hour the sun broke through. I had sensed no clue that the weather would change.

Carolina wren

Over the years this happened to me again and again. But there was more to it, I was to learn. On those misty gray days when I felt certain that rain was imminent, if the Carolina wren arrived and made its short-distance call – again, facing me – substantial rain did not materialize. I learned that I could count on four to six hours without heavy precipitation.

Why does the wren do this? I don't know. I have decided that I don't need to know. I simply need to be grateful for it. I have read accounts of wrens in Africa that seek out humans, squawk at them in a distressed state and then retreat by degrees, coaxing a man or woman to follow. Once when the invitation was taken up, a man found a snake climbing the tree in which the wren had made her nest. It is tempting to interpret this as the wren asking for help, and that may be exactly what was happening. Regardless, the wren does appear to groom a relationship with humans that other birds do not.

Here in Eastern North America, the Carolina wren's contribution to our knowledge of weather can be utilized in a number of ways, one of those being an answer to the question: Which shelter should I construct?

How much time do I have to make a shelter?

By stretching your arm toward the setting sun and turning your flat palm toward your face, you can gauge how much daylight remains in a given day. The distance from the thick bottom of your hand to the knuckle of your thumb approximates

the "path" of the sun taken in one hour. In the mountains, use an imagined hori-zon-line (like a desert's) to calibrate. Smaller hands tend to accompany shorter arms, so the technique proves reliable for all ages.

This technique of reading time is most useful in morning and late afternoon. Mid-day readings are more difficult unless you are certain of the cardinal directions and familiar with the "path" of the sun for a given season.

"My Brother, forgive me for taking your life. Now your pelt will become my second skin, and as I wear it I will honor you. When I eat your flesh it becomes part of my own. I am made of you. Are we not then truly brothers?"

~ the hunter's prayer

CHAPTER 15
Sticks and Stones
~ The Multi-Use Rabbit Stick ~

During the building of shelters in my survival classes, my students are on the lookout for a hardwood stick to be used as a combination shovel/mallet/cudgel/throwing stick. Stick-size will vary from person to person, strength to strength; but, in general, a good throwing stick should be one cubit in length (measured elbow to fingertips) plus the length of the hand added again. For gauging the thickness, wrap thumb and index around the stick. The tip of the thumb should reach to the index fingertip for the stronger, to the first index knuckle for the less strong. A stick too heavy or too light can affect throwing-form.

Smooth the handle (smaller end) to prevent scrapes or cuts when throwing the rabbit stick. Cut or scrape the other end angled on one side to serve as a shovel.

Most right-handed folks who first throw the stick for accuracy miss by throwing to the left. Vice versa for lefties. Throwing the rabbit stick requires its own specialized form. There may be no single correct way to throw, but what follows is a proven, reliable method.

Beyond meat acquisition, the rabbit stick is also a valuable throwing tool for gathering nuts and fruits

from trees and vines. It also serves as a shovel for digging … a mallet for hammering and kneading … and, if curved, a bow in a fire-kit. In a survival situation, you don't know when an opportunity for meat will arise. When it does, you want to be ready. The rabbit stick should be nearby at all times and considered feasible for bringing down only small prey, the range of size varying with the strength of the thrower. Within the scope of this stick's hunting uses are: squirrel, rabbit, bird (grouse or smaller), woodchuck, mink, and muskrat.

The rabbit stick is not meant to kill on contact. It maims by, ideally, injuring a leg or legs … or by stunning. A crippling shot allows the hunter to approach quickly and spare the animal more suffering. The stick can be used as a cudgel to finish off the prey. Crushing the rear of the skull is the quickest and most humane way to kill. The common ethic attached to any throwing stick is to use this method only for survival. Never practice on real animals. Instead, use inanimate targets both on the ground and on the side of a post simulating a tree (a rabbit stick can tear up a tree's inner bark). There are hunting laws governing each state's permissible uses of weapons. Survival methods for hunting are often forgiven only in true life-or-death scenarios!

 ### ***Making a Rabbit Stick*** – Look for a storm-toppled hardwood with a reputation for toughness, such as: oak, hickory, sourwood, dogwood, persimmon, or locust. Locate a proper-sized length of branch, then hack with a stone or knife until it will break off. Smooth a handle on the smaller end for comfort, and then carve or scrape a shovel point on the other end.

Throwing Form

Start throwing the stick as an exercise in form, not strength. For a right-hander, crouch slightly, hold the stick by its handle in your right hand in front of your groin with the far end of the stick pointed straight ahead (as if the stick is part of a bar you are straddling). The left hand can lightly support the stick at mid-length. Place your right foot forward, both legs bent for balance.

Slowly bring the stick back (to the right) along a horizontal plane to the "fully-cocked" position as far as you can point it behind you without rotating your torso. Don't lower your throwing shoulder or raise the stick from its original height. The stick should remain horizontal at all times as if you are sliding it along a tabletop. To achieve "full-cock," the wrist should "break" slightly so that the stick points directly behind you. During this wind-up, keep your chest facing your target. From this "full-cock" position, the stick needs to travel straight ahead on an imaginary straight line from cocked stick to target. In other words, don't let the stick swing outward from the body as it did in the wind-up. The arm has to "break" at the elbow to allow for this perfect line.

While at the full-cock position, begin a slow graceful step forward with the left leg. At the moment before it touches down, the throw begins. On the forward stroke (toward the release), the stick remains parallel to the ground. Release the stick just after the hand passes the torso … not out in front of the body. Bend both legs like shock-absorbers to lower yourself. The right knee may come all the way to the ground for stability, the top of the right foot pressing to the ground, toes pointed behind, the foot sliding on its top surface.

Throwing form: path of the hand

The stick whips forward on that line established by the beginning of the throw. There should be no backlash of hand or wrist to flip the stick. Let it spin on its own. Check your body position after the throw. If your chest is not facing the target, you employed body torque, which will generally bring error into your throw when done at full speed.

Throwing the Rabbit Stick – Following the above form, make a series of easy throws to concentrate on your body position. At this point, don't use a target. Get the form right before you start putting strength into the throw. Few people adapt to this technique right away, so give it a chance to reveal its efficacy.

Coaching a Rabbit Stick Thrower – Explain to a friend the physical requirements of this throwing form. Talk through it as you demonstrate in slow-motion. Then have your new coach choose a different facet to critique each time you throw.

1. Stick should remain parallel to the ground throughout back-swing and fore-swing.
2. Stick should point backwards at fully-cocked position.
3. Stick should move straight ahead from fully-cocked position along one horizontal plane defined by the back-swing. It should not swing out from the body, nor should the end dip toward the ground.
4. Release should be performed just in front of the hip.
5. Chest should remain facing the throwing direction.
6. Shoulders should remain level (no dipping to one side.)
7. Body should move straight forward during throw. (Don't lean away from stick.)
8. Legs should cushion the lowering of the torso at the throw.

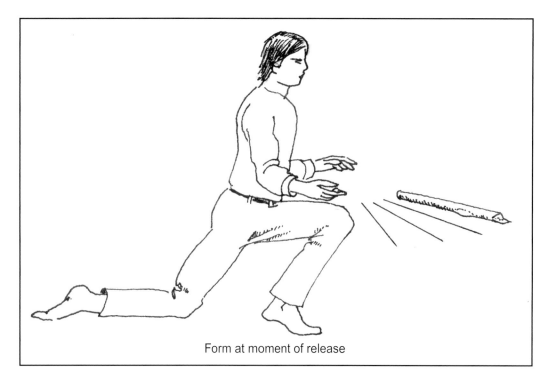

Form at moment of release

Trouble Shooting

The most common breaks of form (> and their repercussions) are these:

1. Holding the stick too long before releasing. > The stick goes wide to the thrower's weak side.

2. Rotating the shoulders. > Same results as #1.

3. Dropping the throwing shoulder and bringing the stick higher like a sidearm toss (right-hander throwing from the right side) with a Frisbee. > Either accuracy or power is sacrificed.

4. Letting the forward-moving stick travel outside the imaginary line connecting the fully-cocked hand to the target. > Same result as #1.

5. Letting the stick hang downward from the arm like a pendulum. > The stick loses its horizontal orientation and either sails high or "walks" along the ground on a skewed plane.

Aiming

When the rabbit stick is flying and spinning through the air, only half of the stick is considered a weapon. For a right-handed throw, the leading edge is always the right half of the stick (at any given moment) as seen by the thrower, because the left half is spinning back toward its point of origin. For this reason, the thrower must aim at the left edge of the target so that the right half of the stick will engage the target.

When throwing at an animal for survival food, consider the animal's feet as the target. By doing so, an error in elevation can pay off with either a direct hit to the body (a throw too high) or a skip off the ground (a throw too low) that hits the animal on the bounce. A direct hit on the feet or legs will likely debilitate the prey.

Throwing Form For a Rabbit Stick Class

– After giving a lesson on form, have each student practice alone for ten minutes, throwing gently without a target. Then gather everyone for critique. Let each student demonstrate throwing while the others point out good or lacking elements of form. With each round, increase throwing speed and power.

Throwing for Accuracy

– Stab a sharpened 1'-high vertical stick into the ground and throw at it from 12 yards. The goal is to hit the branch with the leading half of the throwing stick at ground-level. Gradually pick up speed. Throwing as hard as possible is never necessary. The natural spin of the stick carries plenty of wallop.

The Sacrificial Socks

– Stuff old socks with rags and sew the ankle shut, leaving the sock's ankle-sleeve as a "tail." Tie a tight collar of string (the "neck") at the ball of the foot to define "head" and "torso." Sew two button-eyes on the head. Assemble the throwers and introduce them to the "stuffed sock animals" you have made. Now each student takes a stuffed animal and balances it upon four stick legs impaled into the ground. This small gesture of realism makes the training more personal. The target is now closer to having a "life."

Stalking and Throwing

– Now place the left foot forward (reverse for lefties). In a stealthy ultra-slow-motion movement, stalk the right foot forward. After the right foot alights on the earth, float your weight forward (still in ultra-slow-motion) over this foot and settle there for a time as you stalk the rabbit stick back to the fully-cocked position. This movement should be so slow that the motion is invisible to an observer's peripheral vision. Now stalk the left foot forward into the air. Gather your faculties for the throw. As you purposely lose your balance forward for the final part of this step, commit to the throw. Touch down with the left foot, cushion with your shock-absorbing legs, and throw at your target. (Stalking is covered in *Secrets of The Forest, Volume 3*.)

Sockdolager

– One specially designed stuffed sock goes by the name "Sockdolager." Sew on to him bright button eyes, a dark nose and erect ears. Affix the end of a long spool of light nylon string just to the rear of the top of his head. Under your guidance this fabricated critter can be so life-like that this exercise takes on the intensity of a real hunt.

Beneath a tree limb that is 10' – 15' high, clear a space on the ground for Sockdolager, where he can turn around without getting hung up in leaf litter, grasses, or twigs. Toss the spool over the limb and hoist Sockdolager a few inches off the ground. Keeping the string taut, spin him around a hundred times to twist and tighten the section of string from branch to Sockdolager. Without letting him unwind, lower him to the ground in his clearing. Traction with the ground pre-

vents his unwinding. Keeping tension on the twisted string, walk the spool away as you unreel to a safe place where you will control Sockdolager.

By letting the string go slack he will lean forward and go face-down to "feed." By pulling taut, you stand him up again. By pulling a little more, you'll break his contact with the ground and he will turn by the torque in the twisted string. Though he can turn in only one direction, you can have him face any direction you choose by letting him touch back down on the ground.

Position students in a near half-circle around him at a radius of twenty-five yards. Make sure that you and each student are out of the line of fire if someone were to throw at Sockdolager.

At your signal all stalk toward Sockdolager in ultra-slow-motion. Meanwhile, he is going about his business of eating and periodically looking around, as any respectable prey would. The attentive manner in which you handle his movements brings realism to this exercise.

You are the eyes and ears of Sockdolager. If you consider someone's noise or stalking form to be upsetting, have Sockdolager rise from his feeding position, turn, and stare in that stalker's direction. After a full minute, if all the stalkers in Sockdolager's "range of vision" loyally freeze, have him go about feeding again. But if any stalker gives himself away by a careless movement or sound, call out that stalker's name and count down out-loud from three. (For all other stalkers, time has frozen. They cannot move.) The countdown is, theoretically, the time it takes Sockdolager to react to danger and then bolt. (In this case, three seconds is generous to the hunter.)

string attached just back of the crown of the head so that Sockdolager leans forward to browse when string is slack

sockdolager

The named stalker must throw before "zero," or else he and the throw are disqualified. Whether the throw is a hit or miss – a success or failure – that thrower now becomes a spectator. There is an excellent chance that the thrower will not have his feet positioned correctly, adding a realistic degree of difficulty and variation to the challenge of throwing.

If, during the stalking, a hunter makes a blatantly careless move, call out his name and start the countdown. That's his one chance to throw in the game. If a stalker reaches a tempting distance where he feels he would take a throw in a real hunting scenario, he simply tells you. Again, time freezes for everyone else. Raise Sockdolager to stand with his back to the hunter. This stalker has unlimited time in readying for and performing the throw, but if during this preparation he errs with a hasty move or telltale sound, call his name and begin the countdown.

If a student stalks extremely well and reaches a point that you deem to be the closest reasonable stalk that an animal would allow (clubbing Sockdolager over the head is unrealistic!), tell that stalker that he has concluded a successful stalk

and has unlimited time to make the throw from where he is. Again, if he makes an alarming move or sound, begin the countdown.

You can evoke a lot of personality from Sockdolager by the way you manipulate the string. Because of his bright eyes, his stare is riveting. This exercise comes very close to simulating a hunt. Even people who throw early in the game are mesmerized by the drama as they sit in place and watch the action. The best stalkers who emerge from your group are always last to close in on Sockdolager. They hold a place on center stage longer than the others, and so they become your teaching assistants.

Vertical Throwing

Arboreal animals (perched on the side of a tree) call for another style of throwing that sends the stick spinning in a vertical plane. Now the upper half of the stick becomes the business end, and so the stick must necessarily be thrown at a spot just below the target. In cocking the weapon to throw, bring the slightly-bent elbow directly above the shoulder and hurl the weapon forward with the arm traveling in a vertical plane. Use the same footwork as before.

Arboreal Target – Attach a squirrel-sized sock-target to the side of a post or dead tree. Practice stalking and throwing from varying distances.

The Hare-Trigger! – Set two 5" X 18" logs on end on the ground 20' to 30' apart (depending on the swiftness of your students … faster groups need the longer distance) and then connect the two logs by a length of light rope tied to their tops. The rope should hang in a shallow catenary, as taut as possible without toppling the logs.

One student assumes the role of the rabbit. He kneels on all fours facing the midpoint of the rope, the logs equidistant on his left and right. Behind him a hunter stalks from 25 yards away. Place a handful of carrot slices before the rabbit on a clean surface. He is to remain in place eating, daydreaming, sniffing the air, nibbling occasionally, nodding off, doing whatever rabbits do. If something behind him – a noise, a movement – makes him curious, he cannot turn his body around … just his head to either side, for he must maintain his claim to those carrots.

One hunter/stalker, armed with rabbit stick, approaches at a signal from the teacher. The rabbit settles in, munching on a snack but on full-alert for danger. The other students must meet the challenge of practicing patience, being absolutely quiet as an audience.

When the rabbit eats, the stalker may decide to take advantage of the internal sound of chewing inside the rabbit's head. Becoming bolder, the stalker might make a less cautious move. If a sound disturbs the rabbit, he may stop chewing, freeze, concentrate on listening, and turn his head to one side to take a peripheral peek. The stalker must watch the rabbit at all times to be ready for this – in order to freeze. If the rabbit sees movement or hears an alarming sound, he springs to his feet and dashes to the log of his choice and topples it by pushing away from the arc of rope, thereby, toppling the other log. The stalker must react quickly and throw for the other log before it falls.

If the stalker does an exemplary job, closes the distance, and decides to take a throw, he announces this and the rabbit moves out of harm's way. The rabbit can name which log the stalker must hit. The stalker can take his time and throw. To hit the log is success for the hunter, who wins the carrots. If he misses the log, the game goes to the rabbit.

Before the game begins, it must be clearly understood that the stalker will never throw while the rabbit is at rest … or at the rabbit … or at the log to which the rabbit sprints.

the hare-trigger

"It is no mystery why we keep returning to the river. Water is the stuff of which we are mostly made. Each of us is a tributary, water moving through us at a slow, sustaining pace. When that current stops, we cease to be alive."

~ *Crow Littlejohn*, Requiem to the Silent Stars

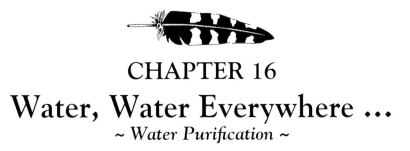

CHAPTER 16
Water, Water Everywhere ...
~ *Water Purification* ~

If the ancient tribes who first lived in North America could return to this world, of all the mind-boggling changes they would see, I believe that the sullied condition of our streams would sadden them more than any other. Each day of their lives these people knelt before a creek and drank the cool lifeblood of the Earth. This practice is no longer feasible. In some areas even the rain falling from the sky is contaminated.

Our bodies depend upon water. Water comprises the greater mass of us. It is the universal solvent through which our nerve cells fire, our blood courses, and our food energy is distributed to the extreme outposts of our body tissue. When the supply runs low every organ of our bodies screams for water. Without it, we shut down. We die.

Think of the role water plays in the miracle of photosynthesis. The solar energy collected by green plants would have no storage closet were it not for water. Nor would plants be able to acquire the nutrients they need from the earth. Yet, this liquid phenomenon is probably the second most taken-for-granted gift on the Earth – the first being the air we breathe. Both are urgent commodities that most people seldom ponder in a day's time.

In rural areas, people who take their water from wells periodically experience a deficit when droughts temporarily deplete water supplies. These people naturally adopt water conservation by utilizing simple practices that may have never occurred to those in urban areas.

In wilderness, a person confronted with the need to find drinkable water might be thrown into the tailspin of dilemma. It is a uniquely unfamiliar problem – one perhaps never before considered. The need is vital. Without it, dehydration

> *After hypothermia, dehydration is our physiological nemesis in a survival situation.*

brings headache, weakness, and disorientation. In short, a person in desperate need of drinking-water loses the ability to take care of himself.

Dehydration leads to complications related to overheating the body and breaking the electrical circuits in the nervous system. After hypothermia, dehydration is our physiological nemesis in a survival situation. If the thirsty person is rash and drinks questionable water, he might find himself in such debilitation that he loses the ability to take care of himself. Drinking water that is contaminated by certain microorganisms can cause prolonged diarrhea, which exacerbates the dehydration problem, leading to more misery, incapacitation, and, ultimately, death.

Six Ways to Obtain Drinkable Water

1. Some trees produce sap that can be drunk by humans. These are the exception to the rule, as most trees make chemicals that render their saps incompatible for human consumption. Maple, most hickories, birch, walnut, and sycamore can be tapped for their sap. (If a taste of hickory sap has a strong medicinal flavor, don't use it.) The results of tapping vary with the seasons – the ideal time being very early spring. From summer on the process will likely be disappointing. Slash a large V-shaped cut (3" along each slope of the V) deep enough into the trunk to cut through both bark layers into the sapwood. At the point of the V, bore an upward-angled, ½"-wide hole into the sapwood (2" in all). Insert a smoothly carved, top-grooved, downward-sloped drip-stick (a crude spile) and collect the drip in a container.

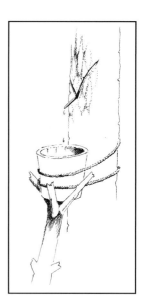

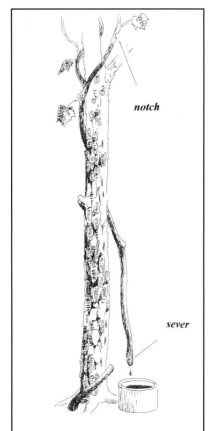

notch

sever

2. In spring and early summer wild grapevine can be severed at ground level, then notched from one side high up on the vine. (Cutting the notch performs the same principle as lifting your finger from the top of a liquid-filled drinking straw.) A steady drip of sweet water can be collected at the ground-level cut.

3. Distillation occurs naturally on those clear nights when the earth cools off enough to condense the humidity in the air. Gather dew drops by mopping wet surfaces with a cloth (a piece of clothing) and wringing it out into a container.

4. A controlled distillation can be performed by a solar still. For this you'll need to be fortunate enough to find a sheet of plastic. Dig a cubed pit about 2'X 2'X 2' into low ground that gets a lot of sun. An open meadow on bottomland makes an excellent site. To increase the quantity of water in the moist pit, tear up green plants and toss them in and pour creek water around the lip of the pit. Set a container in the bottom center of the pit, lay a slack sheet of plastic over the hole and weight down the edges with stones. Now weight down the center of the plastic with a stone so that the sheet assumes the shape of a shallow inverted cone with its point directly over the container. As water evaporates inside the pit, it condenses on the cool plastic. By the cohesive nature of water, molecules form droplets. Gravity leads each droplet to the point of the cone, while the adhesive property of water keeps it clinging to the plastic. When enough droplets coalesce at the low point, their combined weight breaks the bond of the adhesion to the plastic and the container fills. A convenient accessory for this set-up is a length of dried hollow plant stem – like Joe Pye weed or bear paw ("kidney-leaf") rosinweed – to be used as a straw from the container to ground level. On overcast days this is a slow process at best.

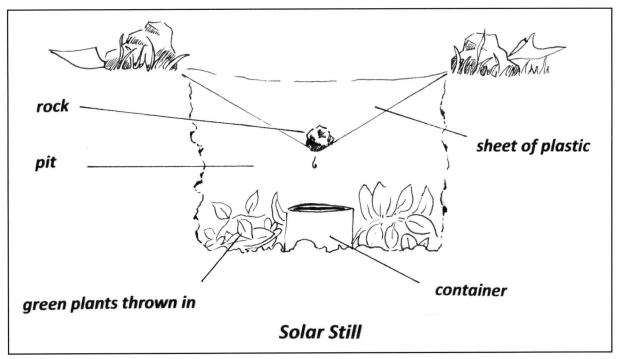

rock

pit

sheet of plastic

green plants thrown in

container

Solar Still

5. A filtering system for creek water can be set up using three tiers of cloth (or "hammocks" hand-crafted from dead bark fibers). See the illustration on the following page. First and highest in the filter-ladder is a layer of shredded/kneaded grass. Sand fills the middle layer and crushed charcoal the lowest. Affix the three cloth hammocks to a tripod of sticks impaled into the ground. A container placed under the three-tiered tower receives the filtered water. Replace the filtering materials after each use. Your source for charcoal is the crispy black char scraped off incompletely burned wood – not the gray ash from a campfire, which is alkaline lye! I have used this filtration method, but I cannot vouch for its efficacy in eliminating pathogens. (I have not had the filtered water tested.) I rely exclusively on the next technique – boiling.

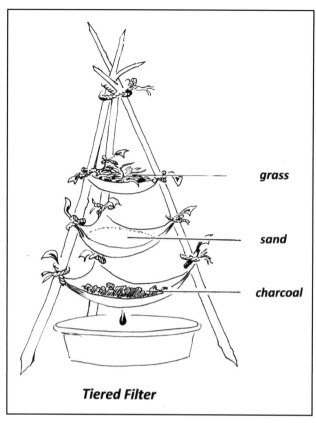

Tiered Filter

grass

sand

charcoal

6. Boiling disinfects. Opinions differ as to how long water should be boiled. The latest research has shown that simply bringing water to a boil suffices in destroying pathogens.

Disinfecting Water by Primitive Boiling

To boil water, you need a container. To make such a container you'll need a chunk of wood, hot coals from a hardwood fire, and a blow tube. The chunk of wood might be a stump that already has a convenient depression for holding water, but it's helpful to have a portable container. Look around the forest for a substantial blank of solid wood from a broken tree.

Anything that funnels an expelled breath into a compressed stream can serve as a blow tube. Natural possibilities are the dry hollow stems of Joe Pye weed, bear paw rosinweed, mature wild lettuce, a section of rivercane … or a segment of any dried (in case of toxicity when green) weed stem reamed of its soft pith (like elderberry).

 ### Making a Blow Tube — If your chosen material is a hard woody substance like rivercane, bamboo or dried elderberry, cut a 7" length using the method described below. If necessary, ream it out with a "ramrod" of thin straight hardwood. Eliminate any sharp points on the ends of the blow tube, for this tube will be partially inside your mouth.

Shortening a Blow Tube without Cracking

If the blow tube material is fragile, like dried Joe Pye Weed, take care not to crush the tube when applying pressure with a knife. First, sharpen your knife. Then make a shallow ringed cut-line (barely a scratch from your blade) around the stem. Hold the stem on the section you plan to keep as a blow tube and cut away from the blow tube by placing the edge of your knife in the scratched groove. Set your holding-hand's thumb on the back of the blade and with that thumb push the blade toward the part of the stem not to be used. As you push, cock back the wrist of the knife hand to coax the blade to make a shallow scalloped cut. Rotate the stem in

your hand and repeat with another scoop right beside the first. Continue circumcising by going all the way around the stem and again. As you patiently cut, allow the stem-excess to break away on its own.

Burning Out a Bowl

From a blazing hardwood fire, remove a hickory-nut-sized coal to burn a cavity into the wood. There are different ways to retrieve coals from a fire, and most are sloppy methods that cause the coal to drop and lose heat. The "chopsticks" method regularly results in a dropped coal. The three-stick method is not much improved unless the sticks are bound together at one end. A shovel-shaped piece of wood or bark can be used as a scoop, but, of course, in time such a tool burns and changes shape and loses its efficiency of design. Even in its prime, such a scoop collects ash as well as coals. Tongs work best.

<u>Making Tongs</u>

<u>Making Tongs</u> - Cut a 2'-long green hickory branch ½"-thick and as straight as you can find. At its midsection, shave 6" flat on two opposing sides. Strive for

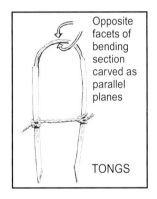

Opposite facets of bending section carved as parallel planes

TONGS

perfectly parallel shaved planes ¼" apart. Heat the midsection over hot coals by constantly turning. When the shaved section is too hot to handle, remove the hickory from the heat, place a shaved side against a 3"- to 4"-thick tree trunk, and bend the stick around it until the prongs are parallel. Let it cool in this position for a minute then shave the ends to flatten the tips for better "pick-up" surfaces. Tie a piece of cordage from one prong to the other midway up the prongs. This cord will be stretched taut when the tongs are at rest.

<u>Making a Boiling Bowl</u>

<u>Making a Boiling Bowl</u> – Saw a 7"-long cylindrical blank from an 8"-thick trunk of dead tulip tree, basswood, buckeye, or other soft hardwood. Set a hot coal on a flat surface of the wooden blank and hold it in place with a green stick as you blow through the tube to superheat the coal. The cavity is slow to start, but the process will speed up. Replace the coal as needed. In time the cavity itself becomes a red coal. When this happens, as you blow, a spurt of blue flame shoots out of the bowl with the roar of a miniature jet engine. It is easy to be hypnotized by this spectacle and overlook the end goal. Take note how your cavity is growing. Change directions in blowing to keep the cavity centered in the wood. If the cavity gets too close to one side, pack mud against that inside wall to prevent a burn-through.

Be careful not to inhale through the blow tube. To prevent hyperventilating, blow slowly. Rest occasionally. Direct all blown sparks safely. Stay upwind to avoid prolonged smoke inhalation. Never stick a sharp blow tube into your mouth!

When the cavity has been achieved, grind creek sand into the bowl using a smooth stone. (Don't feel compelled to remove all the blackened wood. The

charred wood in the bowl is charcoal, a cleansing agent used in the previously mentioned filter system.

Place hickory-nut-sized, non-flaky, non-crumbly stones on the hot coals and, as they heat to incandescence, fill the bowl with creek water. Use tongs to remove a glowing stone from the coals and lower it into the water. Repeat intermittently, removing cooled stones, until the water boils.

 ### *The Bowl-making Cult* – Whenever I tell an adult class that we will spend the evening around our fire blowing on hot coals and making wooden bowls, I always see lackluster expressions on some faces – especially on men, who might consider such a craft to be women's work. Later that evening, after those students get started, I can't get them to put down their blow tubes and go to bed. The project is captivating, even for an observer. It is a witchy scene out of Macbeth to see a ring of *Homo sapiens* breathing fire into wood on a winter's night.

For conducting a class of beginners, cut blanks from a dead "soft hardwood." Provide ½"-thick canes of rivercane or bamboo to be cut into blow tubes. Make a pair of tongs and get started, overseeing the group for safety. This activity will fill an evening. On the next day, hold a water-boiling contest: Who can bring two cups of water to a boil first? (This provides an opportunity to make tea.) The suggested *softer* hardwoods burn out faster, but they will crack in time. However, such soft materials give the novice confidence in completing the chore. All that is needed to replicate the task with harder wood is more time, more coals, and patience.

Ingesting Contaminated Water and its Remedy

Without chemical analysis it is impossible to know what kinds of micro-organisms might exist in a stream – which is why, today, it is never prudent to drink from surface water. As clean as a creek might appear, there could be a carcass rotting in the water … or possibly livestock could be active upstream. The bacteria and other pathogens let loose in the water could cause serious repercussions if drunk. Such illness can take away a person's will to work at the chores necessary for survival. In case of such a sickness, remember sassafras root tea. (Refer to Chapter 5: *7 Plants for 7 Specific Problems*)

sassafras

"Even when circumstances require that there be winners and losers, the transaction is not necessarily a combat."

~ Lewis Thomas, The Lives of a Cell: Notes of a Biology Watcher, *(Penguin Random House)*

CHAPTER 17
Hors D'oeuvres of Protein
~ Adventures with Larvae ~

Virtually every yellow jacket, hornet, and wasp (including immature offspring embedded in cells) dies in winter. Only the queen lives through the winter. So, in an autumn survival scenario, why not eat their latest brood of soon-to-die larvae? At this stage of their growth, these insects provide an important source of protein.

The clever part about eating vegetarian animals is that these animals are innately more knowledgeable than we concerning survival food. They know which plants to eat in order to glean needed amino acids. A typical person in a survival situation does not. In fact, his malnourished body will resort to taking protein from his own tissue by breaking down muscles.

America tends to associate protein with red meat … big slabs of it. But there is protein to be had in tiny creatures. Of the stages that an insect goes through, the larva usually offers the most digestible source of protein, and, conveniently, a larva is easier to catch than an active adult.

Larvae of beetles and ants can be gathered from inside rotten logs, beneath bark, under rocks and logs, and in loose soil. Earthworms can be dug up for food, too. The search can be rather random. Hold any of these caught critters in an improvised corral or other container, where you can feed them flour you have ground from grass seed or pollen. Overnight they will purge their digestive tracts of their present contents. Next day drop them into boiling water or stir-fry with hot coals on a slab of bark for 5 minutes and eat. **All larvae and worms should be cooked to kill parasites!**

A large and satisfying worm-like food of Southern Appalachia is the cicada in its underground phase. This stage (actually the nymph) of the cicada lives in soil for up to 27 years, depending upon the species. This "grub worm" looks like a fat,

white, slightly curled worm about an inch long and as thick as the end of a man's little finger; though they can be found larger.

Gather insect nymphs in creeks under rocks. Collect grasshoppers and crickets from an open field. Remove head, legs, wings, and wing-covers, if present. Catch crayfish in creek shallows and treat like lobster. Boil all these finds and eat. Snake meat is delicious – even a venomous snake. By utilizing these small creatures you are gaining precious *complete* proteins and the fats that your body must have to function properly. (Larger animals will be discussed in the trap and snare section.) With a wealth of plant lore and the acquisition of small creatures, you can survive all but the cold winter months, when you'll need to harvest larger "packages" of protein.

 Starter Larvae – Search beneath rocks and fallen logs and under the bark of dead trees for insect larvae. Collect enough for each student to have a sample. Store the larvae overnight in a container dusted with ground-up grass seed to allow the larvae to consume enough grain to push from their digestive tracts whatever they have recently ingested. Next day, sauté the larvae in a smear of animal fat (or butter) on a cooking rock (see Chapter 18), turning them as you cook. 5 minutes of sizzling heat will suffice. For young students who need to be weaned from their expectations of taste, enhance the flavor of this wild food with seasoned salt.

 Primitive Larva Meal – Boil larvae in the wooden bowl that you created for water-boiling. Eat the larvae without condiments for a better understanding of their true taste. Talk to your students about the importance of protein in their diets and the consequence of not having it. Consider this a "survival meal" and make the shift in diet-attitude that stresses eating not for taste but for nutrition.

Yellow Jacket Stronghold Siege

Yellow jacket soup was a delicacy to the Cherokees. Whenever I mention this to students, they imagine a bowl of dead yellow and black wasps floating in water, but it's the larvae that are the prize. When sautéed in animal fat (or butter), they have a pleasing nutty taste. Many people balk at the idea of eating wormy things and especially at stealing from a nest of notorious stinging wasps, but as you will learn through experience, the mission is not fraught with danger. Of the many nests I have dug up, I have incurred only two stings; however, because the potential is there, you must decide if this harvest is prudent for you, depending upon whether or not you are allergic to stings. An allergic person can die from an anaphylactic reaction.

Finding a Nest

Yellow jackets almost always build underground. The nest, though hidden, looks like a hornet's gray paper nest – the larger-than-football fortress commonly seen perched in winter trees.

Finding a yellow jackets' nest is easy for someone who spends time in summer woods. A subterranean nest has one entrance/exit hole in the ground. It is the size of a fifty-cent piece. Look for intermittent flights incoming and outgoing. Perhaps the most common way a nest is located is when someone is stung multiple times, a sign that a larvae-filled home is nearby. (These wasps generally do not attack *en masse* until their young are implanted in the nest.) If the nest looks like it is embedded in a tangle of tree roots, scratch it off your grocery list and find another less complicated nest site.

A Few Helpful Facts About Yellow Jackets

When a yellow jacket stings, it injects acids and leaves a pheromone on its victim, identifying the "enemy" to other yellow jackets, making the stung person a marked man if he lingers near the nest. Yellow jackets are attracted to certain colors: orange, red and yellow. And whereas they are interested in all sorts of foods that we eat, they do not sting while feeding (unless they are threatened). Note on foods not to carry: The ethylene gas emitted from bananas is similar to the "marked-man" pheromone. As with fire ants, wasps gang-up on a larger enemy for a convincing defense. However, foxes, coyote, skunks, raccoons and bears, all of whom dig up nests to feast on larvae, have a high tolerance for pain, not to mention a protective pile of guard hairs. These predators also have the cunning to perform the deed at night. Yellow jackets are diurnal, sleeping at night. When they are awakened in the dark, their warrior qualities are decidedly dulled.

Early in summer before installing eggs into the cells of the nest, the adults show no signs of being upset at passers-by. I have settled in just inches from a nest-hole in early June to watch the flight patterns. To take that same spectator's seat in August would be a mistake. A yellow jacket's warrior code kicks in once the larvae are implanted inside the comb.

The Harvest

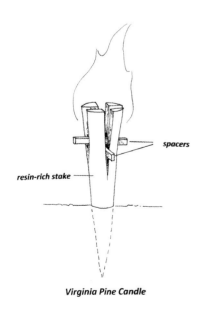

spacers

resin-rich stake

Virginia Pine Candle

Arrive at the nest two hours before dawn with a sharpened stick of lighter-wood (Virginia pine) to ignite and impale into the ground so that the wind carries the acrid smoke over the nest site. This candle is, by and large, an unnecessary precaution but psychologically reassuring for a first dig. For a primitive experience (without a flashlight) the candle serves to illuminate your work site. Quickly and deliberately dig up the nest, break it open, and brush away the groggy adults. Then carry the nest some distance away. Extract larvae and drop them into a cooking vessel, such as a wooden burned-out bowl or a slab of bark on which to stir coals.

Depending upon the developmental stage of the larvae, the skin might be a little chewy. The interior is soft. The taste is pleasantly nutty.

With the coming of daylight the evicted adults become active, but their attention is focused on the old nest site. There they hover about for hours, no doubt, distraught over their loss.

 ### **_Raiding a Yellow Jacket Nest_** – During your first raid, it might seem to go against all instincts to knock on a yellow jacket's door. Timing is all. Attempt this only in late summer or early fall and only during the night. You'll probably want to overdress for protection, but repeated experiences will teach you that such precautions are generally unnecessary.

Use a shovel for your first experience. Expect a football-sized nest and pry it up out of the ground. You'll be pleasantly surprised at the absence of adult activity. This same act in daylight would provoke a firestorm of yellow and black stinging warriors. Break open the nest, evict the sleepy adults, and move at least forty yards from the excavation site to extract the larvae.

 ### **_Making Yellow Jacket Soup_** – On a cooking rock sauté the larvae in animal fat (or butter), turning them so as to heat evenly. Drop the larvae into a boiling bowl, add water, and cook a soup, adding any flavorful herbs you wish – such as wild onion, wild garlic, toothwort root, cattail shoots, or wood sorrel root. Add last any available edible greenery.

Other Wasps

Paper wasps can produce larvae much larger than yellow jackets. Their nests are easy to steal since no digging is required. The security personnel of one nest is not as overwhelming as a hornets' or yellow jackets' nest. Look for paper wasps' nests under dry, natural shelters that are protected from rain. And once again, perform the deed under cover of night.

"He (a Skilloot) ... gave us a roundish root about the size of a small Irish potato which they roasted in the embers until they became soft. This root they call Wap-pa-to ... it has an agreeable taste and answers very well in place of bread. We purchased four bushels ..."

~ William Clark, The Definitive Journals of Lewis and Clark (*Univ. of Nebraska Press*)

CHAPTER 18
A Kitchen in the Forest
~ Cooking in the Wild ~

"Why does everything taste better when it's cooked outside?"

I have heard this question countless times from students when our class is grouped around a campfire and enjoying a primitive meal that we have cooked over flames that they created. The answer probably has a lot to do with the fact that we've just spent an active day outdoors performing more than the daily average workload of the students as they pushed through various projects. But there is more to it. When it comes to the ultimate appreciation of food, there is nothing like fulfilling your dietary needs by your knowledge of the forest, because you know exactly where it came from. You experience an elevated intimacy with food when you meet it on its terms in its natural habitat. When you have spent a good portion of your day foraging and collecting a variety of wild foods for your meal, and when you know that what lies before you "on your plate" is all you will have to eat that night, the concept of appreciation expands with primordial gratitude.

Open Flame Cooking

Most everyone has cooked something over an open flame – even if only a hot-dog on a stick. Little skill is needed. First, know what kind of wood not to use as the skewer – like green sticks of mountain laurel, rhododendron, azalea, mulberry, buckeye, or locust ... all of which are toxic. If your familiarity with trees is limited, learn to identify maple and hickory for use as skewers. Both are safe to use. Simply cut a thin slice of meat, skewer it, and keep it out of the flames as you turn it regularly with patience. Any palatable meat that you harvest can be cooked this way.

Supported Skewer

A technological step up from a hand-held skewer is one that stands on its own and allows you to turn the meat over the flame. It's a simple design. Impale two forked sticks into the ground, one on either side of the fire. The skewer-stick spans the two forks above the fire. Shave any side-branches from the skewer, but leave at least one branch at one end to aid in rotating the skewer.

To ensure that the skewer actually turns the meat rather than rotates inside the meat, use your knife to split the skewer lengthwise for several inches near its mid-point. Sharpen one end of a cross-skewer and pierce the meat through the split in the main skewer and out the other side of the meat. Now two skewers form a cross with their fixed intersection inside the meat.

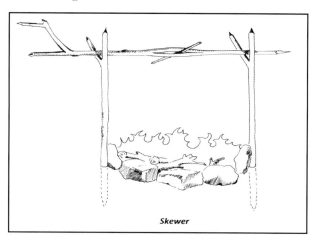

Skewer

 <u>Making a Rotisserie</u> – Cut a three-foot long, ½"-thick straight branch of maple or hickory. At one end leave one or two branchlets sticking out a few inches. This piece will serve as the main skewer. The branchlets will allow you to get a grip on the skewer for turning it. (Or you can fashion a crank handle on this end around the branchlets.) Sharpen the other end of the skewer to a point. Cut a cross-skewer and sharpen one end.

Cut two sticks – each 2 ½' long with a fork on one end. Sharpen the unforked ends and impale them into the ground 6" deep. With the tip of your knife blade, split the main skewer at mid-length for 3" – 4". Stab the sharp end of the main skewer through the meat to be cooked and situate the meat at the midpoint. From a perpendicular angle, stab the cross-skewer into the meat so that its sharp end travels through the split in the main skewer. Prop the main skewer in the two forks and cook over a fire by steadily turning the skewer.

ROCK FRYING

Find a flat, ½" to 1"-thick, fry-pan-sized stone that has neither been underwater nor buried in the earth. (Either could explode when heated.) Set it upon a bed of hot coals by propping the cooking stone an inch above the coals using two sticks of wood. Heat side-A for 10 minutes; turn and heat side-B for 10 minutes; turn again and reheat side-A one last time and then flip to cook on side-A, leaving the rock over the coals. Now you can apply animal fat (or butter) to the rock by simply sizzling the flesh side of a fresh mammal pelt for a few seconds. If your cooking surface loses heat, remove the food, flip the rock, re-oil, and continue cooking on side-B.

Making Pine Bacon – Heat the same cooking rock over coals as described above. From a green branch of white pine, extract a rectangle of inner bark, the size of a slice of bacon. Oil the cooking surface as described above. To win over young students as converts to wild foods, add seasoned salt. Press the strip of bark onto the rock and turn the bark every few minutes until it is crisp enough to snap off when bitten.

THE PIT OVEN

You can bake food underground in a small stone room that has been preheated by fire. The pit oven is a handy cooking site that can take care of many foods at once. Fish, crayfish, strips of meat, and plant greens, roots, and bark can be cooked together.

Cooking a Practice Meal - With a rabbit stick dig a cubicle hole 2' by 2' by 2' and line the bottom and sides with walls of 3"-thick, dense, non-flaky, non-crumbly rocks. Prop long rocks vertically on the pit's sides to reduce the need for stacking numerous rocks. Gathering the oven rocks might take some time, but this is the initiation of involvement for the members of your class. Such a project re-enforces the accountability of wilderness work. *Only those things that we are willing to do are the things that will get done.* These rocks can be used multiple times. Rocks of varied sizes can be used, but none should be smaller than an adult fist. Don't gather stones from a streambed or from underground, as these stones can explode when heated.

Build a blazing fire in the stone box to heat the stones and let it burn for 2 ½ hours, adding fuel as needed. Once the fire is well-established, lay thick logs over the top to reflect heat down to the stones. As a practice for a survival situation, wrap store-bought food (corn, fish, onion, squash, carrots) inside a nest of harvested green grass to serve as a container for the food. If only dead grass is available, wet it thoroughly. Tie the grass bundle with cordage to keep it together.

pit oven

After the stones have heated, scoop out as much ash and coals as possible. Place the grass bundle in the oven (tougher foods on the bottom, softer food on top or in a separate bundle perched above the lower), cap it all with a thick slab of bark that almost spans the pit, and then fill the oven with dirt until you have smothered all spots where wisps of steam or smoke escape. Depending on what is to be cooked, leave the food in the oven 1 to 3 hours. Then dig with wooden scoopers (found around splintered trees), find the bark slab, extract the food, and feast.

 <u>The Real Deal Meal</u> – Collect a variety of wild foods and wrap them in thick swatches of long green grass and/or grape leaves. The wrapping serves to keep the food clean as well as to impart some sweet flavor to the food. Some suggested wild foods to bake/steam this way are: cattail rhizomes, cores of cattail stalks, wapato, groundnut, Solomon's seal tuber, leached acorns, Jerusalem artichoke tuber, larvae, a freshly killed fish or mammal.

Cooking in a Wooden Bowl

Using the same wooden bowl you made for purifying water, you can prepare teas, soups, and semi-solid meals. Heat stones on a bed of hardwood coals, each stone being the size of a hickory nut. Fill the bowl with water and add stones slowly, one at a time, until the water starts to boil. When a stone has lost its heat, place it back on the coals to reheat. Add the food to be boiled and keep the water boiling by periodically adding hot stones.

 <u>Primitive Tea</u> – Inside your wooden bowl, bring water to a boil by adding hot rocks one at a time as needed. Use tongs to handle the rocks. Return cooled rocks to the coals to reheat. When the water has come to a boil, remove all rocks and add tea material to steep.

 <u>Pine Pasta/Beginner Food Boiling</u> – Immerse thin strips of pine inner bark in salt water for three hours. As the bark soaks, heat boiling-stones on a thick bed of coals. When the stones incandesce to red, use them to boil water in your wooden bowl. Rinse the bark in clean water and boil, keeping the water temperature up by adding hot stones as needed. Cook until tender. If you are working with a class of young students, serve the "pine noodles" as you would pasta at home with butter and seasoning or sauce.

A Survival Meal in Autumn

Whenever I give to a class the challenge of devising a way to leach acorns in the creek without losing any slivers of nut, students are often intimidated by the idea of constructing a sieve-basket tight enough to retain the thin slices of acorns, yet porous enough to allow creek water to flow through the tannin-rich nutmeat. The chore might at first seem to require specialized skills, but such is not the case.

A crude basket of grapevine can be constructed that looks like a small, woody basketball net closed at the bottom. Gaps in the basketry can be as open as 2" or 3" square. By stuffing grasses into the basket, you quickly create an effortless sieve.

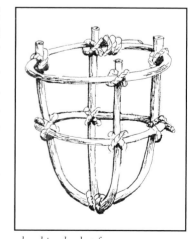

leaching basket frame

Making a Leaching Basket

Making a Leaching Basket – Begin a crude basket with 2 or 3 sections of grapevine, each 2' long, tied together at their midpoints like a woody asterisk. With another 2½' length of vine, make a fixed hoop and secure it by twisting the ends around itself or tying them off. Bend each end of the asterisk vines into the hoop to create a dome shape. The hoop is now the rim of the basket and the asterisk vines make its 4-6 ribs. If pressure from the ribs does not hold them against the rim, tie them off. Add another smaller hoop halfway down the basket, inside or out, and tie it off. If needed, lace smaller vines around the curvature until the basket can hold a mass of green grasses, which will contain the slivers of nuts.

Preparing Acorns for Leaching

Preparing Acorns for Leaching – Set up a nutting stone and hammer stone as described in Chapter 6. Gather acorns (from a white oak, if available). Prepare one acorn at a time by removing the cap, lightly cracking the shell, peeling away the shell, and scraping away any tan, brown, or reddish rind on the outer surface of the nut halves. Place one nut-half flat-side-down on a cutting surface and slice it into the thinnest flakes possible. Repeat until you have enough.

Leaching Tannin from Acorns

Leaching Tannin from Acorns – Fill the basket frame with a mass of grasses and nestle the thin slivers of acorns into the center of the grass. Cover the nestled-in slivers with more grass. Close the basket by jamming a few sticks through the basket just beneath the rim. At a creek submerge the basket on its side with the opening facing upstream. Anchor it with hefty stones on two sides and one on top. Allow to wash overnight.

Completing the Menu

Completing the Menu – After the acorn basket has been anchored in moving water over-night, remove it, and carefully locate the acorns inside the wet grass. The acorns are edible at this point, but they can be rendered more palatable. Into the basket add to the nuts an assortment of other foods: cattail rhizomes and horns, wapato and groundnut tubers. The basket will now serve as your baking container in a pit oven, and the grass will impart an improved flavor to the acorns. Cook one hour.

Making Fruit Leather

Making Fruit Leather – After climbing and shaking the limbs of a fruit-laden persimmon tree in autumn, gather ripe fruit from the ground. It is not true that persimmons require a frost to ripen. The test is simply to assure that a fruit is mushy. If you suspect a fruit is not ripe, a tentative taste-test will apprise you immediately (astringency vs. sweetness). Mash the ripe fruit into a paste, remove the seeds, and spread the pulp across a smooth slab of wood that has been lubricated with animal fat. (For a less primitive class use waxed paper.) Spread the pulp thinly so it will not run when you tilt the pulp toward the fire … but thick enough so that you can embed slices of nuts (leached acorns, pieces of raw walnuts, hickory nuts, or pecans) into the pulp.

If using waxed paper, drape the nutty pulp over the side of a log by making wooden pegs to "nail" the paper to the wood (make slender peg-holes in the log). Build a fire a few feet away and allow the radiant heat to dehydrate the pulp into fruit leather over the next hours. Keep vigil over your leather to avoid scorching when the wind shifts your flame. Be sure to save those persimmon seeds for the delicious hot drink described in Chapter 6.

making persimmon leather

SHELF OVEN

This oven bakes by the steady heat of an ongoing fire beneath a stone cubicle – all of which is built in an earthen pit. It requires arranging flat stones of varying sizes in a design that draws air to the fire and heats an oven compartment above it.

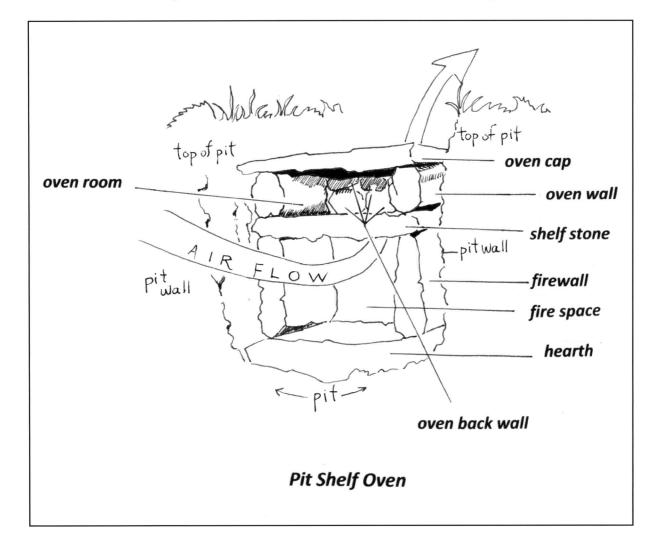

Pit Shelf Oven

Building a Shelf Oven

<u>**Building a Shelf Oven**</u> – Dig a cubical pit 2' - 3' deep measuring 2' X 2' at the sides. Line the bottom with stones to make a mosaic floor or use a large flat stone that covers the bottom. This lowest level makes the *hearth*. On each of two opposite sides of the pit set a flat rectangular stone (or a wall of several stones) on its long edge to create a *fire wall* that rises up half the height of the pit. On the two sides of the pit without fire walls, remove more dirt for a ramp-like appearance for adding fuel.

Spanning the two fire walls, lay an inch-thick flat stone as a *cooking shelf*. (If you use multiple stones for the shelf, fit them closely together so that your food will not be exposed to direct flame. It may be necessary to lay smaller flat stones over the "seams" between shelf stones.)

Upon the shelf stone, lay on edge two rectangular oven wall stones as upward extensions of the fire walls. These stones should rise above the shelf only a few inches. Spanning across the oven wall stones lay an oven cap stone(s) parallel to the shelf and jutting out over the shelf (on both ends) to trap heat rising from the fire. The dimensions of this rock are perhaps the most important, but it can be any thickness. Though not ideal, several long stones could be laid side by side if a large enough cap stone cannot be found, but the seams between them will have to be sealed with slivers of rock or wood. Though you might be tempted to chink the seams with earth, don't. Dried dirt will eventually sprinkle down into the oven compartment.

Once the food is in the oven, build partial walls on the front and back of the oven compartment's cooking shelf. These walls should not reach the cap stone, because heat needs to enter the oven through these openings.

SPICE SUBSTITUTES FOR WILDERNESS COOKING

All-Spice – dried fruit of spicebush (*Lindera benzoin*)

Anise – leaves of sweet goldenrod (*Solidago odora*)

Caraway – seeds of Queen Anne's lace (*Daucus carota*) **(See warning in Chapter 6.)**

Coffee – roots of chicory (*Cichorum intybus*), dandelion (*Taraxacum officinale*), tubers of chufa (*Cyperus esculentus*), seeds of cleavers (*Galium aparine*), persimmon (*Diospyros virginiana*)

Curry – root of lovage (*Ligusticum scothicum*) **Do not confuse with poison hemlock!**

Garlic/onion – wild garlic (*Allium canadense*), wild onion (*Allium cernuum*), wild leek (*Allium triccocum*), root of wood sorrel (*Oxalis montana*)

Lemon – berry of black gum (*Nyssa sylvatica*), greenery of wood sorrel (*Oxalis montana*)

Parsley – leaves of honewort (*Cryptotaenia canadensis*) **(Warning: Some plants with parsley smells are deadly to eat!)**

Pepper – flower bud of day lily (*Hemerocallis fulva*), root of evening primrose (*Oenothera biennis*), seeds of peppergrass (*Lepidium campestre*), leaf of smartweed (*Polygonum hydropiper*) and watercress (*Nasturtium officinale*), root and leaf of toothwort (*Dentaria diphylla*)

Soup thickener (gumbo) – acorn flour, cattail pollen and root starch, curly dock root, greenbrier root, pine pollen and inner bark flour, dried sassafras leaves, violet leaves

Sugar – the boiled-down sap of birch, grapevine, maple, and, to a lesser degree, sycamore and walnut requires too much effort and industry for a survival scenario; better to use dried fruit

"Keep your eyes sharp and be wary. The two-leggeds have learned to hunt us even when they are at home asleep in their beds."

~ *Trickster Rabbit talking at the Great Forest Council to Yonah the Bear*

CHAPTER 19
An Army of Silent Hunters
~ Traps and Snares ~

The number of potential diseases transmitted through wild animals is discouraging, many of them quite serious. Often the method of transfer is through ticks and rodents. Mice – the most frequently killed animals in deadfall traps – are now considered risky for use as a survival food because of their associated pathogens. There is no way for a survivalist to determine that a given mouse is infected with a disease. Ticks only increase the threat. Pathogens can be inhaled, ingested, absorbed by touch, or injected by bite. For your edification, here is a list of maladies to research: Lyme disease, Rocky Mountain spotted fever, Brucellosis, Giardia, Q-Fever, Bubonic Plague (and other plagues), Rabbit Fever (Tularemia), Rabies, Salmonella, Ehrlichiosis, and Hantavirus.

In a true survival situation, you'll have to weigh the potential danger of disease against your need for food. Traps and snares should be utilized only in an emergency. There are state laws that govern and/or restrict their uses. Generally, only licensed trappers can legally set traps, and they are obligated to visit their traps on a regular schedule. A true survival scenario would, of course, preclude the need for a license. Staying alive is more important than a trapping law.

Pathogens aside, the worth of a hunting device that can stand by itself in a remote spot is considerable. It works for you 24 hours a day. It makes no sound or movement until it springs. If care has been taken to mask human scent with aromatic plants, a trap gives no warning to prey.

The lore of traps and snares is more than construction; it's knowing where to set them. This means that your observation of animal signs is important: recognizing runs, feeding areas, trails, scratchings, droppings, gnawings, and bedding grounds. The skills of tracking and stalking prepare you for this and are covered in *Secrets of The Forest, Volume 3.* Your knowledge of plants and your experiences in observing animals will guide you in your choice of baits.

 <u>Scouting for Trap Sites</u> – Walk the periphery of a meadow and look for established trails in the undergrowth that connect forest to meadow. Without lingering at these trails, make a quick search for any spoor that might tell you more about the user; such as, a print or track pattern that reveals gait, hairs, droppings, and size of "tunnels" through grass. Select a place nearby in the trees that would serve as a vantage point for observation. Choose a tree to climb or build a rough blind to conceal yourself.

An hour before twilight, de-scent by rubbing down with aromatic plants, and then stalk into the blind or up into the tree, being sure to avoid the area around the run. Settle in and wait. You'll add to your animal education by connecting the characteristics of the "run" to the creature that uses it. And you will know more about where to place a trap and what size it should be.

Deadfalls

Deadfall traps are set to a baited trigger that releases a heavy weight to fall upon the animal that nibbles on the bait. It takes practice to fine-tune a deadfall. Three designs are covered here. Keep in mind that it is illegal to practice this on live animals. Even if a trapper is licensed, he has a moral obligation to check his traps regularly in case a wounded animal is suffering.

When setting these traps, take precautions not to have the deadfall collapse on your hands while setting a hair trigger device. Working in groups is safer than alone, because one person can hold the deadfall in place while another makes adjustments in the danger zone.

Often called the "Paiute Figure 4," this clever device requires three strong but slender sticks, each with a precision-carved notch and/or wedge-shaped end. Each

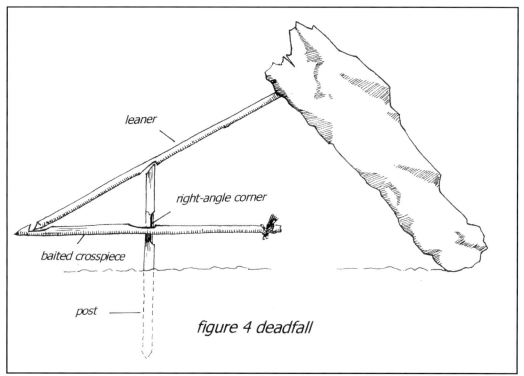

figure 4 deadfall

notch will articulate with another stick's wedge or a squared-off corner, as you will see. Depending upon the size of the falling piece, this trap can harvest chipmunk, squirrel, rabbit, mink, and some birds.

For the falling weight, find a large flattish rock, preferably one that has one end somewhat squared to serve as a reliable hinging interface with the ground. If such a rock is not available, you can shim an uneven side with rock wedges to level the earth-side of the rock when at "full-cock."

Making and Setting Up a Figure 4

Making and Setting Up a Figure 4 – Find a rock that will serve as the deadfall – heavy enough to make an instantaneous kill but light enough to transport. Carry the rock to your trap site, stand it on end, and let it fall. Just beyond the reach of its impact is the spot where you will impale the post (see illustration above) vertically into the ground, deep enough to be stable in the particular soil at the site. Before planting the post, you will have carved a point on the bottom end, a wedge at top, and two flat facets that make a 90° angle as seen in the diagram. The figure 4 can be made any size you prefer. Its limiting factor is the weight of the deadfall and your ability to handle it. Make your first Figure 4 "chipmunk-size." Use a post as thick as your little finger and 14" long – 6" of which will be driven into the ground.

After marking the spot where the post will be driven into the earth, stand the rock upright and brace it so that it cannot fall. Now carve a wedge at the top of the post from two angles, like a steep-sloped, straight-ridged roof. (This ridge will be oriented parallel to the rock's flatness.) Midway up the above-ground part of the post, carve two one-inch-long flat sides that make a right angle. This creates a vertical corner on the midsection of the post that will serve as a catch for the notch that will be carved into the horizontal *bait stick* or *crosspiece*. Drive the post into the ground by hand. Don't hammer it.

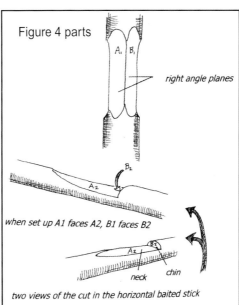

Figure 4 parts

right angle planes

when set up A1 faces A2, B1 faces B2

neck *chin*

two views of the cut in the horizontal baited stick

When set up, the bait stick will form a perfect cross with the post (at the carved corner) and extend into the danger-zone of the leaning rock. Bait will be snugly fastened to the danger-end of the stick with tightly tied cordage. Cut a notch into the bait stick at its point of intersection with the post. This notch should resemble a straight "chin" protruding from the carved "neck." The chin will hook over the corner of the post. The bait-less end of the crosspiece needs another "neck-and-chin" notch turned 90° from its first notch. (This second notch faces skyward.)

On the lower end of the *leaner* stick cut another wedge to fit into the chin-notch on the crosspiece. Where the leaner touches the top of the post, cut a chin-

notch with the chin hooked over the post's wedge. The high end of the leaner will be the prop that holds up the leaning rock. Make certain that when the rock falls it will not hit the post. If necessary, adjust the rock's position.

Assemble the Figure 4 by hooking the crosspiece to the post and applying pressure away from the rock. Holding that in place, hook the leaner like a see-saw over the post (lowering its chin-notch onto the post's wedge) and fitting the lean-er's lower wedge end into the chin-notch on the end of the crosspiece. By applying pressure down on the high end of the leaner you can now release the crosspiece. Ease the rock onto the upper end of the leaner and test your mechanism. If the pressure of the rock tries to create some torque and twists the leaner to one side rather than holding it in its original orientation, the trap will trigger immediately. You'll have to alter the rock's position on the ground. A slight tug on the bait end of the crosspiece should bring the rock down with lethal effect. Be careful of flying parts when deadfalls are tripped during practice.

The Figure 4 Team Construction Project – After demonstrating a

working Figure 4 and after drawing detailed diagrams of post, crosspiece, and lean-er, divide your class into teams of 3. Search the forest for deadfall rocks and straight sticks of dead, dry wood. After a lesson on safe knife use, ask each team to make a trap. Each team member has one stick to shape and contribute to the finished trap.

With the projects complete, tour the traps for a show-and-tell. See which de-vice is easiest to trigger and study its crosspiece-to-post connection to understand why that particular carving of notch and corner make the collapse more fluid.

An Improved Trigger for the Paiute Deadfall
(By John and Geri McPherson, Prairie Wolf Publishing, Randolph, Kansas)

The previous, historical design of the Figure 4 (by the Paiutes) teaches precision carving skills, which is a big part of the value of the project. After whittling out the requisite angles and notches on the three wooden pieces (and after painstakingly figuring out how to get the parts to integrate with one another and actually balance the rock at full-cock) you will appreciate the improvement of the trigger system that follows. For this design virtually no knife-work is needed. A strong, Y-shaped stick (the **Y-post** fulcrum) is impaled into the ground at an angle. Another strong stick will serve as a lopsided **seesaw** over the Y. The short end of the seesaw sup-ports the edge of the rock (deadfall), while the long end is held in place by a length of cordage that makes a half-turn around the post of the Y. A short, rounded **stub** of stick is tied to the other end of the cordage. A slender **bait-stick** locks the stub in place, setting the trap at full-cock.

The Prairie Wolf Paiute Deadfall – Temporarily prop up a deadfall

rock vertically to set the stage for your construction. Just outside the fall of the rock, impale a sharpened fulcrum (Y-post) firmly into the earth at an approximate 75°-80° (leaning toward the deadfall rock). Be sure that the rock, when it falls, will

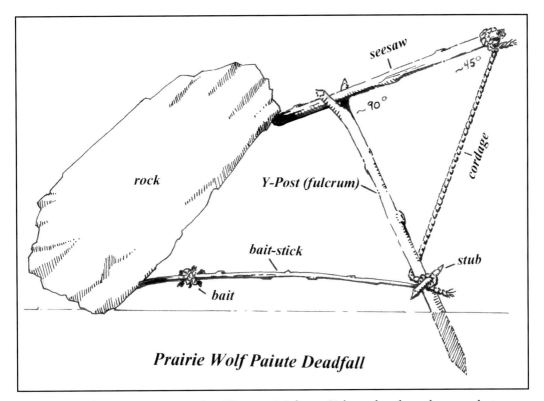

Prairie Wolf Paiute Deadfall

come close but not contact this Y-post. Make a 2' length of cordage and tie one end to a 10"-long seesaw stick. Hold the seesaw in the Y-crotch with 2" of the seesaw (the rope-less end) protruding from the Y toward the rock. Angle the seesaw slightly so that it lies perpendicular to the post of the fulcrum. Extend the loose end of the rope to the post 2" above the ground. Wrap the rope a half-turn around the post to mark the place on the rope where you will tie the 2"-long stub. Now tie the rope to the center of the stub and set up the parts again to test your measurements. You'll want ~90° at the seesaw/fulcrum angle and ~45° at the seesaw/rope angle. After making any needed adjustments by retying the stub, hold the seesaw in place while you wrap the stub around the post.

Jam a temporary "Y-twig" into the ground to hold the stub in place. (See illustration below.) Apply pressure downward with one hand on the short end of the seesaw. With the other hand gently lower the rock to rest on the short end of the seesaw. Replace the temporary Y-twig with a long, slender bait-stick (not a Y) that merely pushes against the free end of the stub and keeps it from unwinding on the post. The other end of the bait-stick can be jammed into the underside of the rock.

Bait should be tied to the bait-stick near the rock. When the prey gnaws on the bait, the disturbance jostles the bait-stick from the stub, which can fly around the post to loosen all the tension on the seesaw, allowing the rock to fall. When triggering the trap during practice, be careful of the seesaw and the whip of the stub as they fly through the air.

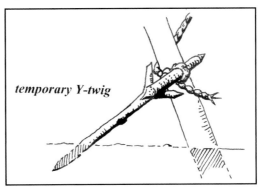

temporary Y-twig

The Whip-Trigger Deadfall
(By John and Geri McPherson)

This deadfall uses a dead dry stick (4' to 5' long) that bends like a bow and, when released of its tension, whips away the prop stick that holds up the rock. Prop the rock with a temporary brace so that you can work safely in the kill-zone. Next to the rock inside the kill-zone, dig a half-walnut-size crater. Into the center of this crater hammer a 1'-long, 3/8"-wide rounded trigger post vertically into the ground until the top of this post sinks just below the rim of the crater.

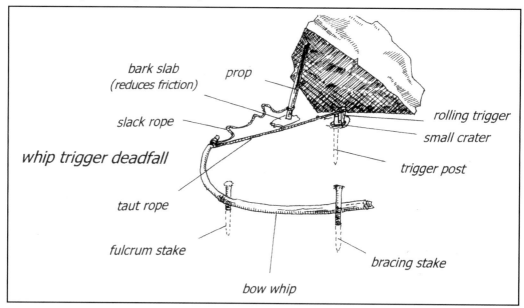

Outside the kill-zone, firmly hammer two 1 ½' stout *bow stakes* into the ground (all but 6") to anchor the bow. One is the bracing stake on the backside of the bow near its far end; the other is the *fulcrum stake* on the belly of the bow at midsection. The bow will bend toward the rock. (Standing trees, if available, can be used in place of one or both stakes.) Tie a length of cordage (the *taut rope*) to the whipping end of the bow. Tie bait to the taut rope just inches from the crater, being sure that the bait is positioned well inside the kill-zone.

Now carve a *rolling trigger*, which looks like a rounded, 1½ inch-long, <u>sharpened</u> #2 pencil with a groove encircling the eraser end. Tie the free end of the taut rope firmly into this groove.

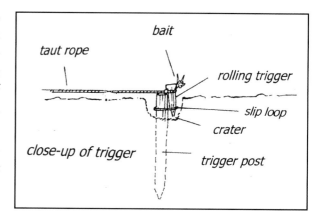

Make a *loop* of string slightly larger than the circumference of the **trigger post** and loop it over the post. Hold the bow bent with your foot and position the rolling trigger on the opposite side of the post from the bow, with the taut rope running <u>over the top</u> of the post. Now raise the loop up the

post enough to include the tapered end of the rolling trigger. Holding the trigger and loop in place, replace your foot pressure on the bow with your free hand and carefully let the bow take the tension of the taut rope. The pressure of the <u>rolling trigger</u> against the <u>post</u> keeps the bow bent and holds the trap cocked for action. The rounded circumference of each of these two parts results in a hair-trigger. All that is needed now is to connect the deadfall to the action.

Tie a second rope (*slack rope*) from the whipping end of the bow to the low end of the *prop* and now set the prop to support the leaning rock. Discard the temporary rock brace. Be sure to leave the slack rope several inches too long. This slack gives the bow unimpeded freedom to whip for a few inches before jerking the *prop* free. To make certain that the prop can be pulled out, set its low end on a smooth platform of bark or wood that has been inlaid flush with the ground; otherwise, the prop is likely to impale the ground from the weight of the rock and get stuck there.

The trap is now fully-cocked. A nudge on the bait disturbs the fragile balance of the two rounded surfaces of *rolling trigger* and *post*. When the rolling trigger moves off-center, it slips and releases the bow, which jerks free the *prop*. Down comes the rock. If the prop balks at being pulled, adjust the length of the taut rope to bend the bow more severely for a more forceful tug.

Making The Whip-trigger Deadfall

– Collect all the wooden materials needed. For a first effort, supply your students with store-bought nylon string. (Later, for the pure primitive experience, have them use handmade cordage.) Each time you cut the nylon, singe the cut ends to prevent raveling. (**Take care not to drip melted nylon onto your skin!**) As before, form teams to complete traps. Tour the finished traps for product testing. The trickiest aspect of this trap is getting the rolling trigger and trigger post to articulate. Make sure that the post is vertical. For young students: If a particular trap proves to be virtually impossible to "cock" due to a slippery connection, *lightly* shave a very shallow flattened surface on either trigger or post.

The Step-on Platform Trigger
by Gordon Nagorski, Ontario, Canada

For this "treadle" trap, a solid framework is first erected by hammering two *vertical posts* into the ground 3' apart. Drive two longer *diagonal posts* into the ground and lash each to a vertical post (making facing triangles) to leave two X-cradles at top. Laid into these cradles, a *crossbar* completes the frame. This frame will support a heavy log that, when triggered, falls just outside the frame next to the vertical posts. The *trigger* is a platform of sticks (*treadle*) propped upon a bar raised off the ground by a rope connected to a higher *falling bar*. Refer to the illustration as you proceed.

Making the Step-on Platform Trigger Deadfall

– Hammer into the ground two stout sticks – 3' *vertical posts* – positioned 2 ½' apart. Two longer sticks (*diagonal support posts*) are hammered at an angle, each one crossing a ver-

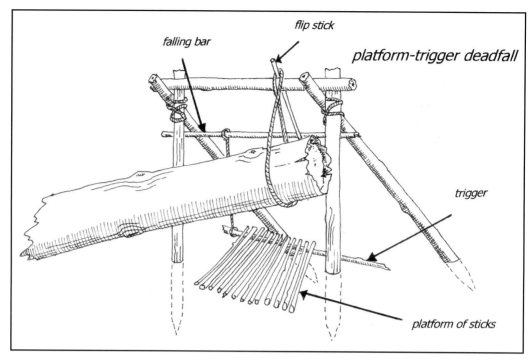

falling bar

flip stick

platform-trigger deadfall

trigger

platform of sticks

tical post to make a lopsided X at its top. This begins the trap's frame as two right triangles facing one another in mirror image, each with a *cradle* at the top. Lash each vertical post to its diagonal support at the point of their intersection. (To better secure the lashing, pre-cut notches where the wood pieces join.)

For a first-time practice on this trap design, give your students a supply of nylon string. Later, for the pure primitive experience, have them use handmade cordage.

Set a stout *crossbar* from one cradle to the other. It need not be lashed. Another horizontal – the *falling bar* – will stretch from one hypotenuse to the other *inside the triangles* several inches below the top crossbar. A strong pencil-thick stick – the *flip stick* – is broken to size to reach from the inside of the falling bar to the top of the crossbar with 2" sticking above the crossbar. Lots of pressure downward (against the crossbar) on this projecting 2" of the flip stick can hold the falling bar in place against the diagonals. The *deadfall* supplies this pressure.

Find a 6'-8'-long, heavy log at least 10" thick to serve as the deadfall. Tie a loose *loop* (several inches too big) of cordage around the log. Lift the log high enough to carry the loop over the two-inch projection of flip stick above the crossbar.

Though you'll be wishing for a third hand, the setting of this trap can be accomplished solo with the help of a bent knee in holding up the log. Your hands will be busy, both applying pressure to the flip stick (to hold the falling bar in place) and looping the string over the flip stick. Adjustments to the size of loop may be necessary.

The trap is now at "full-cock" but with no way to trigger it other than pulling down the falling bar. **Be careful as you work around the raised log!** Tie another string to the middle of the falling bar and tie the other end of the string to the

raised end of the **trigger** or **platform bar**. That raised end of the trigger should hover off the ground several inches.

Now lay down a series of 1 ½'-long **platform sticks** inside the framework perpendicular to the platform bar, one end of each stick on the ground and one end on the bar. This makes a *platform* which, if stepped on, pulls down the falling bar, frees the lower end of the flip stick (**Watch your eyes! Beware of the flip stick!**) and allows the log to crash down.

Bait goes on the high side of the platform. You'll want the animal to approach the low side of the platform and step on it; therefore, any other approach must be discouraged by obstacles placed in the way. Erect walls of "stockade" sticks to channel the animal to the desired entry. The falling log is intended to break the backs of medium-sized prey such as woodchuck or raccoon.

Teams Constructing the Platform Deadfall – As before, form
teams of 3 and supply your students with nylon string. The pieces of the support frame are best crafted by hatchet, but only experienced adults should wield this tool. The most challenging part of this construction is finding the proper deadfall, heavy enough to be lethal but light enough to carry. It may be necessary to access deadfalls by cutting sections of a dead tree with a saw. **To secure the log during set-up, provide a temporary safety prop to prevent an accident!**

The Spring Snare

Find a recently expired, standing 2"-3"-thick sapling to use as a **spring**. (Living trees lose their spring.) Carefully bend the tree over so that it does not break, prune away limbs, and tie cordage high up on the trunk. On the loose end of this cord, just above ground level, attach a *J-shaped snare hook* carved from dead wood (See illustration.) Tie this cord into a groove carved near the top of the "J."

Carve an *inverted J hook stake* to be hammered into the earth as a fixed post.

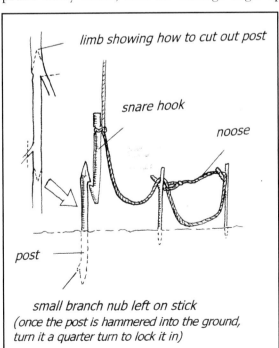

limb showing how to cut out post

snare hook

noose

post

small branch nub left on stick
(once the post is hammered into the ground,
turn it a quarter turn to lock it in)

Making the Spring Snare
– For safety's sake build this one as a group. The danger of having the tree whip upward unexpectedly could result in the snare hook snagging someone. Have several students hold the tree in its bent position while others work with the

string and J-hooks. Switch roles so that everyone gets experience setting the J hooks. For an adult working alone, setting the snare is easy.

Fine tune the interface of the two J-hooks so that they work as a smooth trigger. The top snare hook should easily slip off the inverted hook.

With the snare "uncocked", tie another cord to the snare hook to make a small lasso, whose noose can be propped open by forked twigs pinned into the ground. Bait is not necessary if this snare is set up on an animal run. Consider the animal you expect to snare (rabbit, raccoon, squirrel) and adjust the noose so that the animal's muzzle could nose through the loop. As

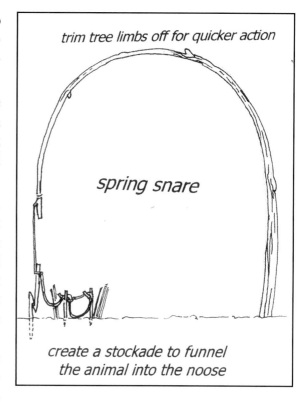

trim tree limbs off for quicker action

spring snare

*create a stockade to funnel
the animal into the noose*

the prey proceeds through the noose, the lasso closes around the thicker part of the animal's torso. The top J hook is tugged off the lower one and the spring is sprung. The tree whips upward and suspends the prey in its noose.

The animal is not killed, nor is it usually injured. Within earshot, you'll hear it squeal. Get to the snare, lower the tree, and dispatch the animal quickly.

There is a reason that a run is well-established. Animals are creatures of habit. The challenge with this design is finding a suitable tree near the run. A live sapling may feel springy for a few practice snaps, but it will eventually lose its spring after being bent too long. Would it be possible to bring a dead tree to your preferred snare site? Could you set it up at your place of choice so that it serves as a spring? Think outside the box. How could you do that?

Bringing the Spring to the Snare – Consider the bow you used in the Whip-trigger Deadfall. Instead of hammering bow stakes into the earth, use the branches of a living tree (or cordage tied to the trunk) as brace and fulcrum. This turns the whip-trigger mechanism 90°, springing the bow vertically instead of horizontally. Whereas the former bow did not require lashing to stakes, this bow needs to be secured to tree limbs.

You may be fortunate to find a low dead flexible branch still attached to a living tree. This can serve as a bow and there will be no need for a brace and fulcrum. Or there may be no trees near your building site. In that case, you'll need to hammer a long stake into the ground and then pull it free, so that you can use the long slender hole in the ground to insert your chosen spring.

Final Notes on Traps and Snares

These devices have no conscience or discretion. Domestic dogs and cats and even children could be hurt by them. Do not leave your practice models loaded. Even though using these traps in practice is illegal except in dire emergency, it is a good idea to learn to make and set them up. Should you ever need these traps, your familiarity with the process will boost your confidence.

Bibliography

BOOKS

American Indian Medicine by Virgil Vogel; University of Oklahoma

American Indian Plants by Charles Millspaugh; Dover

Edible Wild Plants by Elias & Dykeman; Sterling

Edible Wild Plants by Lee Peterson; Houghton Mifflin

Edible Wild Plants by Samuel Thayer; Forager's Harvest

Edible Wild Plants of Eastern North America by M. Fernald & A. Kinsey; Dover

Field Guide to the Ferns by Lloyd Snyder & James Bruce; University of Georgia

A Field Guide to Medicinal Plants by Arnold & Connie Krochmal; Times

Fruit Key and Twig Key to Trees and Shrubs by William M. Harlow, Ph.D., Dover Press

Grasses by Lauren Brown; Houghton Mifflin

The Herbalist by Joseph Meyer; Meyerbooks

Identification of Southeastern Trees in Winter by Richard Preston/Valerie Wright; NC State

Indian Herbalogy of North America by Alma Hutchens; Shambhala

Magic and Medicine of Plants; Reader's Digest

Medicinal Plants (Eastern/Central) by Steven Foster/James Duke; Houghton Mifflin

Medicinal Plants of Southern Appalachia by Patricia Kyritsi Howell

Myths and Sacred Formulas of the Cherokees by James Mooney; Cherokee Heritage

Naked into the Wilderness: Primitive Wilderness Living & Survival Skills by John & Geri McPherson; Prairie Wolf

Native Trees of the Southeast by Brown, Kirkman & Leopold; Timber

Nature's Garden by Samuel Thayer; Forager's Harvest

Nature's Healing Arts by Lonnelle Aikman; National Geographic

Newcomb's Wildflower Guide by Lawrence Newcomb; Little, Brown

Outdoor Survival Skills by Larry Dean Olsen; Pocket Books

Poisonous Plants of the United States and Canada by John Kingsbury; Prentice Hall

Primitive Technology, A Book of Earth Skills; Society of Primitive Technology; Gibbs Smith

Stalking the Wild Asparagus by Euell Gibbons; Alan C. Hood

This Green World by Rutherford Platt; Dodd, Mead & Company

The Uses of Wild Plants by Frank Tozer; Green Man

Want Natural Colour? by Jeanie Reagan; self-published

What a Plant Knows by Daniel Chamovitz; Scientific American, FS&G

Weeds in Winter by Lauren Brown; Norton

Wild Foods Field Guide and Cookbook by Billy Joe Tatum; Workman

Winter Botany by William Trelease; Dover

Winter Tree Finder; Nature Study Guild

MAGAZINES and PERIODICALS:

Bulletin of Primitive Technology; P.O. Box 905, Rexburg, ID 83440

The Forager; Wild Food Institute, P.O. Box 156, Port Wing, WI 54865

The Wild Foods Forum; Eco Images, P.O. Box 61413, Virginia Beach, VA 23466

Wilderness Way, Primitive Skills and Earth Wisdom; P.O. Box 621, Bellaire, TX 77402

INDEX

Numbers in bold indicate plant photographs

Gourdy Old Style on 100gsm (70#) matte artpaper
Case 14pt. C1S with gloss film lamination – Flexi-Bind
Type and design by Karen Paul Stone